LIGHT METAL ALLOYS APPLICATIONS

Edited by **Waldemar A. Monteiro**

Light Metal Alloys Applications
http://dx.doi.org/10.5772/57069
Edited by Waldemar A. Monteiro

Contributors
Waldemar Alfredo Monteiro, Zuzanka Trojanova, Tomasz Arkadiusz Tański, Anna Dobrzanska-Danikiewicz, Krzysztof Labisz, Masafumi Noda, Hisashi Mori, Yoshio Gonda, Kunio Funami, Kenji Fujino, Sebastian Feliu Batlle, A Samaniego, Ricardo Ambriz, David Jaramillo, Victor Songmene, Seyed Ali Niknam, Yasser Zedan, Rajat Roy, Robert Schneider, Burkhard Heine, Richard J. Grant, Diego Manfredi, Matteo Pavese, Flaviana Calignano, Elisa Ambrosio, Sara Biamino, Riccardo Canali, Manickavasagam Krishnan, Paolo Fino

Published by InTech
Janeza Trdine 9, 51000 Rijeka, Croatia

Publishing Process Manager Iva Lipovic
Technical Editor InTech DTP team
Cover InTech Design team

First published June, 2014
Printed in Croatia

Additional hard copies can be obtained from orders@intechopen.com

Light Metal Alloys Applications, Edited by Waldemar A. Monteiro
p. cm.
ISBN 978-953-51-1588-5

For EU product safety concerns:
IN TECH d.o.o., Prolaz Marije Krucifikse Kozulić 3, 51000 Rijeka, Croatia,
info@intechopen.com or visit our website at intechopen.com.

Meet the editor

Waldemar Alfredo Monteiro is a Physicist, MSc (Solid State Physics), DSc (Nuclear Technology) at the University of São Paulo, São Paulo, Brazil. He is a lecturer on undergraduate and graduate courses at Presbyterian Mackenzie University. He is also a senior researcher at Nuclear and Energy Research Institute (Materials Science and Technology Center) – University of São Paulo. His expertise areas are Physical Metallurgy, Powder Metallurgy and Materials Characterization (Optical and Electron Microscopy; Microanalysis Techniques). Dr. Monteiro has published 2 books, 4 book chapters and 150 articles (scientific journals or congress proceedings). The academic advisories include 40 scientific initiation (undergraduate students), 25 MSc and 14 DSc.

Contents

Preface

Light alloys have become of great importance in engineering for construction of transportation equipment. Many of these light weight alloys have sufficiently high strength to warrant their use for structural purposes, and as a result of their use, the total weight of transportation equipment has been considerably decreased. One of the greatest applications of light metals is in aircraft construction. At present, the metals that serve as the base of the principal light alloys are aluminum and magnesium.

One of the most important lightweight alloys in use for several applications are the aluminum alloys, a good example is in the aerospace and aeronautical segment, accounting for 80% of total amount of the weight in commercial aircraft nowadays, which are employed for structural components, parts and plates. In fact, the development of aluminum alloys follows the aircraft industry from the beginnings of its history early in the 20th century. For structural components, commonly, wrought aluminum alloys parts could be found in forged, machined and assembled parts. However, some casting parts that have low cost of production also play an important role in aircraft parts. Aluminum alloys also have an important part in spacecraft, since they have been used in components and equipment as brackets, fixtures, chassis, covers and casing for many tools and devices.

Magnesium is among the lightest of all metals, and is also the sixth most abundant metal on earth. Magnesium is ductile and the most machinable of all metals. Magnesium alloy developments have traditionally been driven by requirements for lightweight materials to operate under increasingly demanding conditions. This has been a major factor in the extensive use of magnesium alloy castings, wrought products and powder metallurgy components. The biggest potential market for magnesium alloys is in the automotive industry. Other applications include electronics, sporting goods, office equipment, nuclear applications, flares, sacrificial anodes for the protection of other metals, flash photography and tools, biocompatibility of magnesium alloys in orthopedic and cardiovascular applications, modern foundry techniques for magnesium alloys, powder metallurgy processing applied to magnesium alloys.

Waldemar Alfredo Monteiro
Materials Science and Technology Center
Nuclear and Energy Research Institute,
São Paulo, Brazil

School of Engineering
Presbyterian Mackenzie University,
São Paulo, Brazil

Section 1

Aluminium Alloys

Chapter 1

Additive Manufacturing of Al Alloys and Aluminium Matrix Composites (AMCs)

Diego Manfredi, Flaviana Calignano,
Manickavasagam Krishnan, Riccardo Canali,
Elisa Paola Ambrosio, Sara Biamino, Daniele Ugues,
Matteo Pavese and Paolo Fino

Additional information is available at the end of the chapter

http://dx.doi.org/10.5772/58534

1. Introduction

In a product development context, the term Rapid Prototyping (RP) was widely used to describe technologies which created physical prototypes directly from digital data. The first methods for rapid prototyping became available in the late 1980s and were used to produce models and prototype parts. Users of RP technology have come to realize that this term is inadequate to describe the more recent applications of these technologies. The ASTM F-42 committee was recently formed to standardize Additive Manufacturing (AM) terminology and develop industry standards. According to their first standard, ASTM F2792-10, AM is defined as *"The process of joining materials to make objects from 3D model data, usually layer upon layer, as opposed to subtractive manufacturing technologies"*. The basic principle of this technology is that a geometric model, initially generated using three-dimensional Computer Aided Design (3D CAD) system (e.g. CATIA, Pro/Engineer, SolidWorks), can be manufactured directly without the need of process planning [1].

There are many related terms used to describe AM and common synonyms include: additive fabrication, additive layer manufacturing, direct digital manufacturing, 3D printing and freeform fabrication. Within the last 20 years, AM has evolved from simple 3D printers used for rapid prototyping in non-structural resins to sophisticated rapid manufacturing systems that can be used to create functional parts in different engineering materials directly without the use of tooling. Most work to date has been conducted using polymer materials, but the development of AM processes such as Selective Laser Sintering/Melting, Electron Beam

Melting and Laser Engineered Net Shaping enabled to build parts by using metallic materials, metal matrix composites and ceramic materials. Additive manufactured parts are now utilized in aerospace, automotive, medical fields and also in consumer products and military [1-3].

Additive manufacturing or 3D printing is receiving unprecedented attention from the mainstream media, investment community, and national governments around the world. This attention reached a pinnacle when 3D printing was mentioned by United States President Barack Obama in his February 2013 State of the Union address. AM, just 25 years old and still a relatively small industry, has completed a transformation from obscurity to something that many can't stop talking about. In 2011, about 1600 articles were found. In 2012, they were 16000, a tenfold increase. Prototyping has been the technology's biggest application, thus the name rapid prototyping, and it remains a key category. The fastest-growing application, however, is in the actual manufacturing of parts for final products. In just 10 years, this important application has grown from almost nothing to more than 28% of the total global product and service revenues. The manufacturing of final parts, rather than prototyping, is where the manufacturing money is, and it is the most significant part of AM's future. Researchers and industry leaders in the European Union (EU) have identified AM as a key emerging technology. Teaming relationships have been formed between university, industry, and government entities within and across countries. The overall level of activity and infrastructure in the EU is greater to that of the U.S. in this key area. Several large cooperative projects have been funded, worth of millions of euros across Europe, among them AMAZE (Additive Manufacturing Aiming Towards Zero Waste & Efficient Production of High-Tech Metal Products) [4], RAPOLAC (Rapid Production of Large Aerospace Components) [5], Custom Fit (aiming at mass customized consumer and medical project manufacturing) [6], E-BREAK (Engine Breakthrough Components and Subsystems) [7], TiAlCharger (Titanium Aluminide Turbochargers-Improved Fuel Economy, Reduced Emissions) [8], and many others. Large aerospace companies, such as Boeing, GE Aviation, and Airbus, are hard at work qualifying AM processes and materials for flight. Boeing, for example, now has 200 different AM part numbers on 10 production platforms, including both military and commercial jets [9].

In 2013, NASA and Aerojet Rocketdyne have been testing rocket engine components made using additive manufacturing. At temperatures approaching 3300 °C, the AM rocket fuel injectors performed identically to conventionally manufactured parts. The series of tests demonstrated the ability to design, manufacture and test a critical rocket engine component using selective laser melting (SLM) manufacturing technology. This type of injector manufactured with traditional processes would take more than a year to make, but with these new processes it can be produced in less than four months, with a 70 percent reduction in cost [10]. The General Electric Aviation has used SLM to produce parts for its upcoming LEAP (Leading Edge Aviation Propulsion) family of turbofan engines, made in conjunction with France's Snecma.

AM technologies have the potential to change the paradigm for manufacturing, away from mass production in large factories with dedicated tooling and with high costs, to a world of mass customization and distributed manufacture. AM can be used anywhere in the product life cycle from pre-production prototypes to full scale production, as well as for tooling

http://dx.doi.org/10.5772/58534

applications or post production repair. AM processes are stimulating innovation in component design, enabling the manufacture of parts that cannot be made by traditional methods and are stimulating alternative business models and supply chain approaches. AM enables the manufacture of highly complex shapes with very few geometric limitations compared to traditional manufacturing processes.

This freedom of design, for example, has led to the technology being used to manufacture topologically optimized shapes with improved strength to weight ratios, a fundamental aspect in both aerospace and automotive design to reduce vehicle weight. For the bio-medical industry, AM is already leading to a revolution in customized medicine where dental implants, orthopedics, and hearing aids are manufactured to fit an individual's unique physiology.

However, despite many examples of growth and progress, many myths and misconceptions associated with the technology have developed:

- AM is a low-labor content "pushbutton" technology.

At the moment, a lot of work occurs before and after the actual production of the parts: in fact data needs to be prepared at the front end, which may require CAD expertise, the repair of the models, and optimization of support structures. Moreover, main building parameters, such as layer thickness, temperature, scan speed, and a number of other options may need to be adjusted for a particular part or type of material.

- AM is fast.

As the name implies, AM systems build parts by depositing, fusing, curing, or laminating consecutive layers of material. These layers are typically 0.025–0.250 mm in thickness, so parts often require thousands of layers. Therefore the preparation of large parts could take even several days.

- AM systems can produce anything.

Most of AM systems can successfully build shapes that cannot be fabricated easily or at all using conventional methods of manufacturing. However, AM processes also have limitations. One is minimum wall thickness. Another is the requirement for supports on down-facing surfaces, which can be difficult to remove. Material that is trapped in internal channels can also be difficult or impossible to remove, and the dimension of the internal channels determines the degree of difficulty in removing material.

- AM systems and materials are inexpensive.

Generally, industrial AM systems are more expensive than CNC machining centers, and starting materials are far more expensive (for example high quality powders, with controlled chemical composition, size and shape).

In any case, as AM continues to develop on multiple levels and in many directions, it will gain more momentum and respect as a method of manufacturing. Metal parts from some AM systems are already on par with their cast or wrought counterparts. As organizations qualify and certify these and other materials and processes, the industry will grow very large. 3D printing is not competing with conventional manufacturing techniques, but is instead complementing and hybridizing with them to make new things possible.

2. Additive manufacturing processes for metals

It was less than 10 years ago that these systems began to gain traction commercially. Today, seven companies, all in Europe, offer systems based on powder bed fusion technology. Examples of powder bed fusion are direct metal laser sintering (DMLS) from EOS and selective laser melting (SLM) from Renishaw. The most popular metal systems use a laser to heat and melt fine particles in a powder bed. Parts obtained with these systems are impressive, especially the fine features and small internal channels that are possible.

Systems that instead use an electron beam as the energy source are much faster, but the surface finish and feature detail are not as impressive. In the case of some orthopedic implants, however, a rough surface is preferred. This, coupled with speed, is why many medical implant manufacturers have selected electron beam melting (EBM) systems from Arcam (Arcam is currently the only manufacturer to use electron beam energy in a powder bed). Many metal parts that are currently being made by AM can be cost prohibitive, difficult, or impossible to produce using conventional methods, such as casting or machining.

This gives AM an advantage, but only if the right types of parts are selected. Quality requirements play a role. Good feature detail and surface finish are possible and comparable to metal castings, but they do not match the surface quality of CNC machined parts. If a part can be produced conventionally at a reasonable cost and the volume is relatively high, it is often best to go that route at the present time. The metal AM systems are complemented by extensive CNC machining and other traditional processes.

Many types of metals are available on metal AM systems. Among the most popular for medical and aerospace applications is the titanium alloy Ti-6Al-4V. Other metals used are cobalt–chrome, stainless steels, tool steels, aluminium alloys such as AlSi10Mg, jewelry and dental gold alloys, nickel-based superalloys such as Inconel 625 and 718, and TiAl alloys. Aerospace-grade aluminium and other metals are in development. All of the unused metal powder can be recycled in the machines after sieving. Industry standards are becoming increasingly important as companies apply AM to the production of final products. In January 2012, ASTM International Committee F42 on Additive Manufacturing Technologies approved F2924-12 Standard Specification for Additive Manufacturing Titanium-6Aluminium-4Vanadium with Powder Bed Fusion. It is the first AM material standard by ASTM and it could give a boost to the additive manufacture of Ti-6Al-4V.

The technology classification agreed by the AM SIG (Special Interest Group) can be seen detailed in Table 1, which shows seven top level classifications for additive manufacturing technologies, below which there are a range of different material classifications, and discrete manufacturing technologies produced by a range of global companies.

http://dx.doi.org/10.5772/58534

Classification	Technology	Description	Materials	Developers (Country)
Binder Jetting	3D Printing Ink-jetting S-Print M-Print	Creates objects by depositing a binding agent to join powdered material	Metal, Polymer, Ceramic	ExOne (US) VoxelJet (Germany) 3D Systems (US)
Direct Energy Deposition	Direct Metal Deposition Laser Deposition Laser Consolidation Electron Beam Direct Melting	Builds parts by using focused thermal energy to fuse materials as they are deposited on a substrate	Metal: powder and wire	DM3D (US) NRC-IMI (Canada) Irepa Laser (France) Trumpf (Germany) Sciaky (US)
Material Extrusion	Fused Deposition Modeling	Creates objects by dispensing material through a nozzle to build layers	Polymer	Stratasys (US) Delta micro Factory (China) 3D Systems (US)
Material Jetting	Polyject Ink-jetting Thermojet	Builds parts by depositing small droplets of build material, which are then cured by exposure to light	Photopolymer, Wax	Stratasys (US) LUXeXcel (Netherlands) 3D Systems (US)
Powder Bed Fusion	Direct Metal Laser Sintering Selective Laser Melting Electron Beam Melting Selection Laser Sintering	Creates objects by using thermal energy to fuse regions of a powder bed	Metal, Polymer, Ceramic	EOS (Germany) Renishaw (UK) Phenix Systems (France) Matsuura Machinery (Japan) ARCAM (Sweden) 3D Systems (US)
Sheet Lamination	Ultrasonic Consolidation Laminated Object Manufacture	Builds parts by trimming sheets of material and binding them together in layers	Hybrids, Metallic, Ceramic	Fabrisonic (US) CAM-LEM (US)
VAT Photopolymerisation	Stereolithography Digital Light Processing	Builds parts by using light to selectively cure layers of material in a vat of photopolymer	Photopolymer, Ceramic	3D Systems (US) EnvisionTEC (Germany) DWS Srl (Italy) Lithoz (Austria)

Table 1. Classification of additive manufacturing processes defined by the AM SIG – adapted from ASTM AM classifications (2012) [11].

3. Direct metal laser sintering — DMLS

Direct Metal Laser Sintering (DMLS), a trademark of EOS GmbH (Germany), is an Additive Manufacturing (AM) technology that creates parts in a layer-by-layer fashion directly from computer-aided design by selectively fusing and consolidating thin layers of powders with a scanning laser beam. DMLS is a net-shape process, producing parts with high accuracy and detail resolution, good surface quality and excellent mechanical properties. As mentioned before, it has many benefits over traditional techniques and its application for manufacturing three dimensional objects represents one of the promising directions to solve challenging industrial problems. Moreover, since the components are built layer by layer, it is possible to design internal features and passages that could not be cast or otherwise machined. Complex geometries and assemblies with multiple components can be simplified to fewer parts with a more cost effective assembly. DMLS does not require special tooling like castings, so it is convenient for short production runs. Applications using this technology include direct parts for a variety of industries including automotive and other industries that use complex parts of small to medium size. For instance, concerning space applications, when the "buy to fly ratio" of a structure is below about 30%, meaning that more than 70% is machined away, additive manufacturing technologies become an alternative to conventional machining for complex metallic parts. And when direct manufacturing allows to produce in one shot spacecraft structures resulting from complex assemblies, then the offered possibilities are really huge. DMLS has also already been used to fabricate lightweight structures, similar to cellular structures. Inspiring to natural systems, a variety of synthetic cellular solids could be made including stochastic foams, consolidated powders, hollow sphere structures, honeycombs, textile laminates and lattice block or miniature truss structures. The cellular metal structures have been used in various industrial applications such as heat exchangers, in automotive and aerospace industries, thanks to their valuable characteristics as low density, high strength, good energy absorption and good thermal properties. Virtually any material can be produced by additive manufacturing, most of those by DMLS, with the exception of some intermetallic and high temperature alloys.

One of the most interesting types of these materials are aluminium alloys. It is rather easy to produce alloys with compositions suitable for casting, since they will melt easily, but recently wrought alloys have also been fabricated by this technique, and new ones are being studied. The microstructure of these alloys is peculiar, since the very fast cooling occurring after the melting induced by laser provides an ultra-fine microstructure and this is responsible for a significant increase of the mechanical properties that can be obtained with conventional casting processes followed by heat treatments. It is also possible, to increase specific properties of the alloy, to produce aluminium matrix composites (AMC). These could have applications in high-demanding components, for instance in automotive and aerospace industries, where it is necessary to improve stiffness, hardness and high temperature properties. Ceramic discontinuous reinforcements are the most suitable reinforcement that can be envisaged.

4. DMLS process

Initially, all contour of the layer structure is exposed with a selected laser power and contour speed. As the diameter of the melted zone is usually larger than the laser diameter, it is necessary to compensate the dimensional error and the laser beam must be shifted by half the

http://dx.doi.org/10.5772/58534

width from the contour to the inside, to make sure that the contour of the later part will correspond exactly to the original CAD data. This correction of the position is called Beam Offset (BO) [12]. The BO value is again defined with respect to the edge of the boundary (Figure 1), and if this value is higher or less than the correct value, the particles of the irradiated region may be not melted or over-melted. During hatching, the laser beam moves line after line several times to assure that the melting process can unroll completely. The distance between the lines is called hatching distance (h_d) and is set about one quarter of the laser beam.

Another important parameter that can lead to a distorted part or a process interruption is the layer thickness. If the value is too high, no optimal adhesion between the single layers can be realized because the melting depth is not high enough. Furthermore, mechanical tension can be generated through this layer which can lead to detachment of the layer below. If the selected value is too small, a tearing-off of a structure can happen during the recoating process, since the melted particles get struck between it and recoater blade [12].

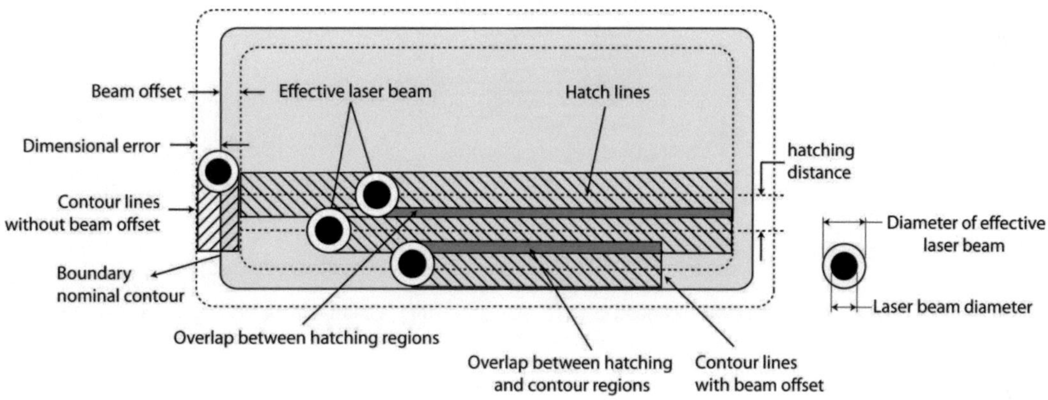

Figure 1. Exposure strategies.

5. DMLS scanning strategy

Many scanning options are provided in commercial SLM machines which include skywriting and hatch pattern along *x*, *y*, both in *xy* and alternating in *xy* for different requirements. DMLS EOS machine has up-skin and down-skin options to improve mechanical properties by allowing user to assign different process parameters at adjoining layers. Subsequent paragraphs illustrate salient features of scanning strategies available with DMLS process.

5.1. Skywriting

During scanning a certain time is needed to accelerate the mirrors to the desired speed. This is due to inertia of mirrors used for scanning. During this time laser beam cover some distance in which speed is not constant hence more energy is applied at the edges of the part than the inside of the part. To avoid this situation and to keep acceleration and retardation phase out of the layer geometry, skywriting option is provided in DMLS machines. If skywriting option is selected, then the mirror is accelerated already before the start of the part so that it has reached the desired speed before the beginning of exposure. Laser is switched on at the start of the part. Similarly retardation phase begins at the part end where the laser is switched off as shown in Figure 2.

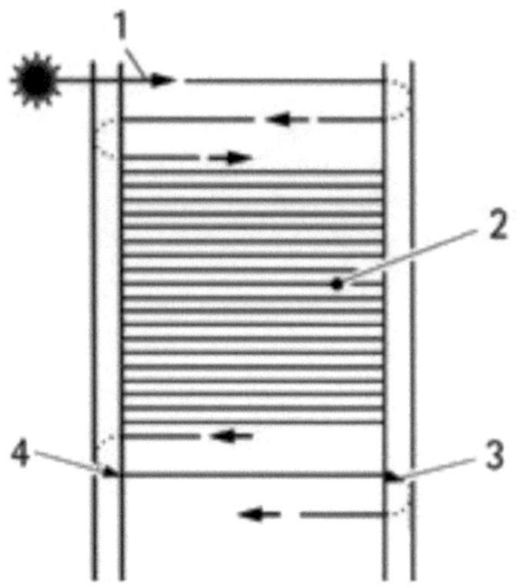

Figure 2. The exposure vector is shown by a solid line (1). The dashed line presents the skywriting. In this area the laser beam is run with a laser power of 0 W (3). With the beginning of the part (4), the laser power switched on with a sized value and the laser beam is running with a constant velocity.

5.2. Hatch pattern

Four choices for hatch pattern selection are generally available, i.e., along *x*, along *y*, both in *xy* or alternating in *xy* as shown in Figure 3. Scanning can be done either along *x* or along *y* (Figure 3a and 3b). If both in *x and y* options are selected than there will be double exposure on the layer, once along *x* and then along *y* (Figure 3c). In alternating in *xy* choice, direction of scanning is changed for alternating layers (Figure 3d). Figure 3e shows the direction of scanning rotated of 67° between consecutive layers. This is default value of the hatch pattern in DMLS EOS machine.

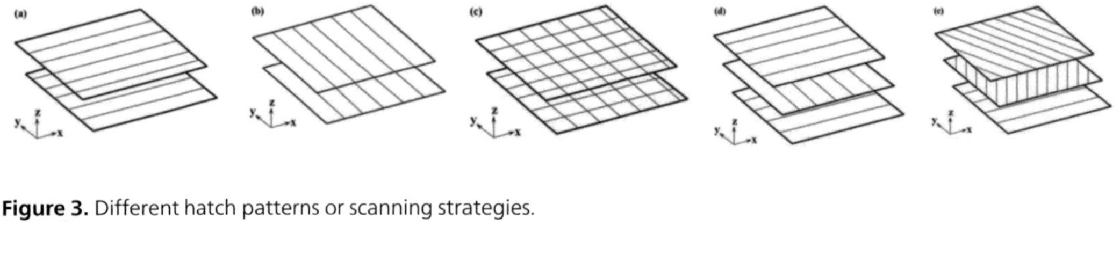

Figure 3. Different hatch patterns or scanning strategies.

http://dx.doi.org/10.5772/58534

5.3. Up-skin and down-skin

The layer above which there is no area to be exposed, is termed up-skin, and is built in three layers. Similarly the layer below which there is no exposed area is called down-skin, and is built in two layers. The rest of the part is termed core or in-skin, and is shown in Figure 4. Commercial machine software calculates up-skin, down-skin and core areas and different parameters can be assigned for these areas. Some overlapping area between core and up-down-skin area can also be assigned for improved joining among these zones.

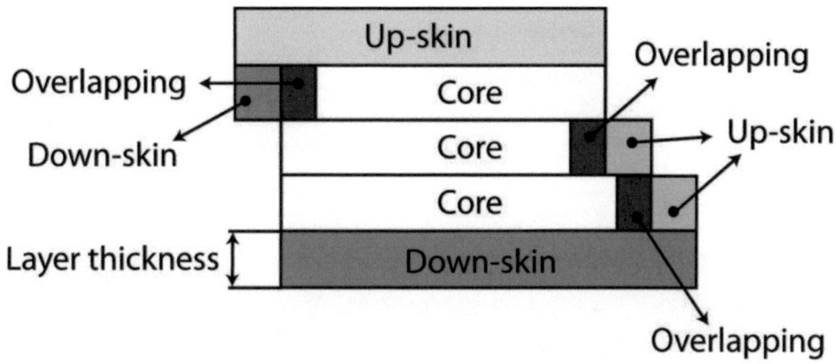

Figure 4. Up-skin, down-skin and core regions.

6. Materials processed by DMLS

Table 2 condenses a list of metal-based materials processed by SLM/DMLS as available in the literature.

Some of the main common features for different materials processed by SLM are summarized hereafter.

Microstructure: small grains, non-equilibrium phases and new chemical compounds are formed during SLM of metal powders [28,40].

Mechanical properties: due to the formation of very fine grains and non-equilibrium structures, the SLM parts could exhibit better mechanical properties (yield strength, ultimate tensile strength and ductility) than the wrought products [28]. Yasa et al. [20] have studied the Charpy impact test of TiAl6V4, A316L stainless steel and maraging steel parts produced by SLM. Spierings et al. [41] have investigated the influence of three different size distribution of 316L stainless steel powder on mechanical properties and surface quality. Their research concluded that powder with smaller size particle distribution could be easily melted and yields high density, high mechanical strength and productivity.

Anisotropy: the parts are produced by stacking layers in SLM, due to this reason the parts have some anisotropy in a particular direction. Buchbinder et al. [42] have investigated the fatigue

Family	Material	References
Metals and alloys	Steel	hot-work steel [13]
		stainless steel 316L [14-17]
		martensitic steel [18]
		tool steel [15-19]
		maraging steel [20]
	Titanium	Ti6Al4V [21-24]
		Ti6Al7Nb [25]
	Nickel based alloy	Inconel 718 [26]
		Inconel 625 [27]
	Copper	copper [28-29]
	Gold	Gold [30]
	Aluminium	A6061 [31]
		AlSi12 and AlSi12Mg [32,33]
		AlSi10Mg [34-37]
Composites	MMC	Fe-graphite, Ti- graphite/diamond, Ti-SiC, AlSi-SiC, AlMg-SiC, Co-WC, Fe-SiC and Cu, Ni, Ti, C, Cu-TiC and Cu, Ni, Ti, B_4C, Cu-TiB_2 [38]
		Al4.5Cu3Mg-SiC [39]
	CMC	ZrO2, Y_2O_3, Al, Al_2O_3 and TiO_2,Al,C TiC/Al_2O_3 [38]

Table 2. Examples of materials used in SLM/DMLS.

testing of AlSi10Mg parts produced by SLM. Their investigation concluded that sample produced in x plane exhibited higher fatigue resistance than samples produced in other orientations. Yadroitsev et al. [27] have studied the mechanical properties of SLM parts. They found that Young's modulus of Inconel 625 parts produced in xy plane are 1.5 greater than that of parts produced in z direction. Qiu et al. [43] studied the microstructure and tensile properties of Ti6Al4V parts produced by SLM in two different orientations. They found that Ti6Al4V parts show high tensile strength, but the Ti6Al4V samples oriented along z axis showed higher ductility than that of sample oriented in xy plane. This anisotropy was caused by the orientation of the columnar grains with respect to the tensile test direction. Thijs et al. [44] studied about SLM of tantalum parts. Their study revealed that grains were oriented in <111> direction, because of partial re-melting of the previous layer, competitive growth mechanism and specific global direction of heat flow during SLM of tantalum. This texture formation had a large influence on yield strength during compression test of tantalum parts.

Residual stresses: these are induced in the SLM parts due to the complexity in melting behavior of layers. Re-melting and solidifying of previously melted layers can cause stress field to exceed the yield strength of the material, causing distortion of the part. Studies show that island scan strategy could reduce the residual stress of SLM parts [45]. Shiomi et al. [46] measured residual stress of chromium molybdenum steel parts produced by SLM. They found that: performing stress relieving treatment of build platform along with parts at about 700 °C for one hour, decreased residual stress by 70%, re-scanning of each layer reduced the residual stresses by 55% and heating of build platform to 160 °C reduced the residual stresses by 40%. Mercelis and Kruth have used crack compliance method and XRD to measure residual stress of 316L stainless steel parts produced by SLM. Zaeh and Branner [47] have carried out Finite Element Analysis to evaluate residual stress of steels processed by SLM and they also measured residual stress by using neutron diffraction.

Surface Roughness: the surface roughness of the SLM parts is higher than that of conventionally produced parts. Vertical faces of the part have lower surface roughness than that of curved or inclined or top faces. To improve the surface finish of the SLM parts, secondary operations like shot peening, machining or abrasive jet machining are carried out. Kruth et al. [48] showed that the surface quality of Stainless steel 316L and Ti6Al4V parts could also be improved in SLM by using Selective Laser Erosion (SLE) and re-melting. Calignano et al. [12] have improved the roughness of AlSi10Mg parts produced by DMLS process.

7. Case study — AlSi10Mg parts by DMLS

7.1. Experimental equipment

Aluminium alloy samples were produced by DMLS with an EOSINT M270 Xtended version. In this machine a powerful Yb (Ytterbium) fiber laser system in an argon atmosphere is used to melt powders with a continuous power up to 200 W and a spot size of 100 μm. The aluminium powder alloy used in this study is a gas atomized one produced by EOS Gmbh (Germany), and its nominal composition is reported in Table 3.

Element	Si	Fe	Cu	Mn	Mg	Zn	Ti	Al
Weight %	9-11	≤0.55	≤0.05	≤0.45	0.2-0.45	≤0.1	≤0.15	remainder

Table 3. Nominal composition of EOS AlSi10Mg alloy powder in accordance with standard DIN EN 1706:2010-06.

This alloy is similar to an A360.2 alloy [49]: thanks to its near eutectic composition in the Al-Si phase diagram it is often used in casting, having a melting temperature around 570 °C. This alloy offers good strength and hardness and is therefore used for parts with thin walls and complex geometry subjected to high loads, as in the aerospace and automotive industries.

Table 4 shows the default values assigned to up-skin, down-skin and core for the scan speed, laser power and hatching distance parameters in producing the Al alloy specimens.

Samples of rectangular shape and 50 × 10 × 3 mm size were produced to analyze the density, hardness and Young's Modulus. Considering tensile tests, specimens were built according to

Parameters	Core	Up-skin	Down-skin	Contour
Scan speed (mm/s)	800	1000	900	900
Laser power (W)	195	195	190	80
Hatching distance (mm)	0.17	0.2	0.1	

Table 4. Default values of scan speed, laser power and hatching distance for EOS M270 Xtended.

the standard ASTM E8M. For the analysis on the surface roughness, the three input or control factors chosen for up-skin were scan speed, laser power and hatching distance (Table 5). The surface finish of the samples was then analyzed though a 3D scanner ATOS Compact Scan 2M (GOM GmbH) and a MarSurf M 300C (Mahr GmbH) mobile roughness measuring instrument before and after post-processing. To compare the different DMLS samples, the average value of the ordinates from centerline, defined as Ra, was used. It is theoretically derived as the arithmetic average value of departure of the profile from the mean line along a sampling length. The shot-peening process was performed with glass microspheres using a sand-blasting machine, SD9 Northblast, in order to improve their surface finishing. The samples, before and after shot peening, were also characterized by a Field Emission Scanning Electron Microscope (FESEM) Zeiss SupraTM 40.

Variable Parameters	Values
Scan speed (mm/s)	800, 850, 900, 950, 1000, 1250
Laser power (W)	120, 155, 190
Hatching distance (mm)	0.10, 0.15, 0.20
Fixed Parameters	**Values**
Layer thickness (μm)	30
Spot size (mm)	0.1

Table 5. Process parameters values for the up-skin.

7.2. Powder analysis

The AlSi10Mg powder was gas atomized and supplied by EOS GmbH. A preliminary observation was made to investigate the morphology of the AlSi10Mg powder particles by using the FESEM described above.

The spherical morphology and smooth surface of the powder particles result in a good flowability and homogeneous layer distribution [36]. Figure 5 shows the FESEM micrograph of AlSi10Mg powder in as received condition. The size of AlSi10Mg powder ranges from 1 to 35 μm with an average size around 23 μm. Bigger clusters of dimension larger than 40 μm were also observed. Thus the powder was sieved to separate the agglomerated powder particles each time before producing the final part in DMLS process.

http://dx.doi.org/10.5772/58534

Material	Orientation	Yield Strength [MPa]	Ultimate Tensile Strength [MPa]	Elongation at break [%]
AlSi10Mg after DMLS	xy-plane	240 ± 8	330 ± 4	6.2 ± 0.4
	z axis	230 ± 5	328 ± 4	4.1 ± 0.3
A360.0 F *	–	170	317	5

Table 7. Mean values of tensile properties of aluminium alloy DMLS specimens produced according the standard ASTM E8M, compared to a similar alloy in as-fabricated conditions [*].

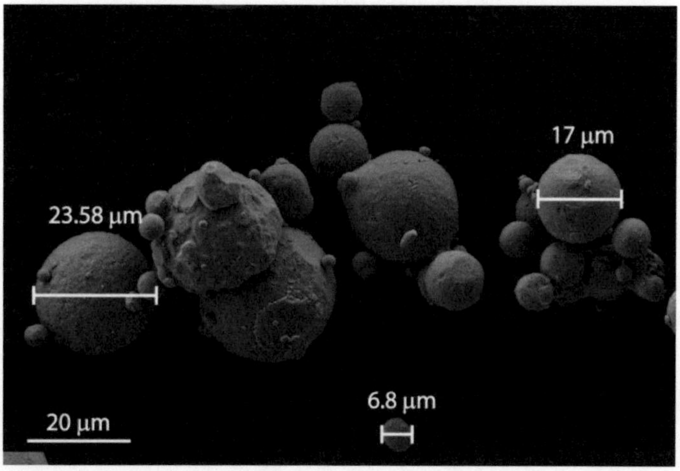

Figure 5. FESEM observation of AlSi10Mg powder.

In Table 6 are summarized the mean values for density, Young's modulus and Vickers Hardness. Moreover, considering for the AlSiMg alloy a theoretical density of 2.68 g/cm^3 [49], the percentage of the residual porosity can be calculated.

Density (g/cm^3)	Residual Porosity (%)	Hardness (HV)	Young's Modulus E (GPa)
2.66	0.8	105 ± 5	73 ± 1

Table 6. Density, Hardness and Young's Modulus of aluminium alloy DMLS specimens.

Considering tensile tests, the results are summarized in Table 7. Variations were not found among performances of samples with different orientations on the powder deposition plane, while there are some differences with the values obtained along the direction perpendicular to it, as already discussed in a previous paragraph. In fact, as the name implies, the parts are produced by stacking layers in SLM, due to this reason the parts have some anisotropy in the so called "building direction", or z axis.

The AlSi10Mg DMLS specimen microstructure was analyzed by optical microscopy (Leica DMI 5000 M optical microscope) and by FESEM. Top and a lateral surfaces of the specimens were polished by using SiC abrasive papers of different grits and then by using 6 μm, 3 μm and 1 μm diamond pastes respectively. Then they were etched with Weck's reagent or HF in water to highlight the scanning tracks or melt pools. The dimensions of the melt pools (width and depth) depend on the laser power and hatching distance employed, and also on the scanning strategy (Figure 3).

Figure 6 shows the optical microscopy images of AlSi10Mg specimen. Hauser et al. have reported that the cross-section of the single tracks formed in SLM process were either crescent shape or elliptical section [50]. From the optical micrograph of the AlSi10Mg sample polished in the lateral surface (z direction), a typical crescent shape structure can be noticed, if the scan track or melt pool is oriented perpendicular to the polished surface (Figure 6, on the left). Some elongated scan tracks are also observed, when the scan track is aligned at an angle with respect to the polished surface. This is due to the scanning strategy of the laser beam adopted (see Figure 3e). Figure 6 shows also the optical microscopy image of AlSi10Mg specimen polished in xy plane. The cross section of scan tracks of different layers and the overlapping of scan tracks can be observed.

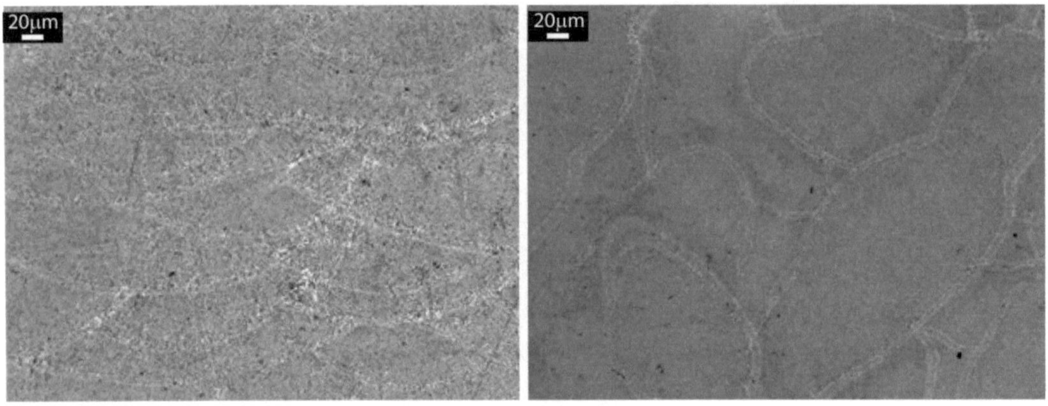

Figure 6. Microstructure of AlSi10Mg along *Z* direction (on the left) and along *XY* plane (on the right) after etching with HF in water reagent.

The geometry of the scan tracks was not clear with the default scanning strategy, because the scan lines are always rotated by 67° with respect to the scan lines of previous layer (see Figure 3e). In order to observe the geometry of the scan tracks, a specimen was produced by orienting the scan lines along x axis. Figure 7 (a) and (b) show the microstructure of AlSi10Mg sample produced by orienting scan tracks only along x axis (see Figure 3b).

It can be observed that the shape of scan tracks is crescent or elliptical. However, this specimen had a higher porosity with respect to the ones obtained with the default scanning strategy. Therefore, the default scanning strategy with 67° is generally chosen to produce AlSi10Mg samples for the measurement of mechanical properties.

http://dx.doi.org/10.5772/58534

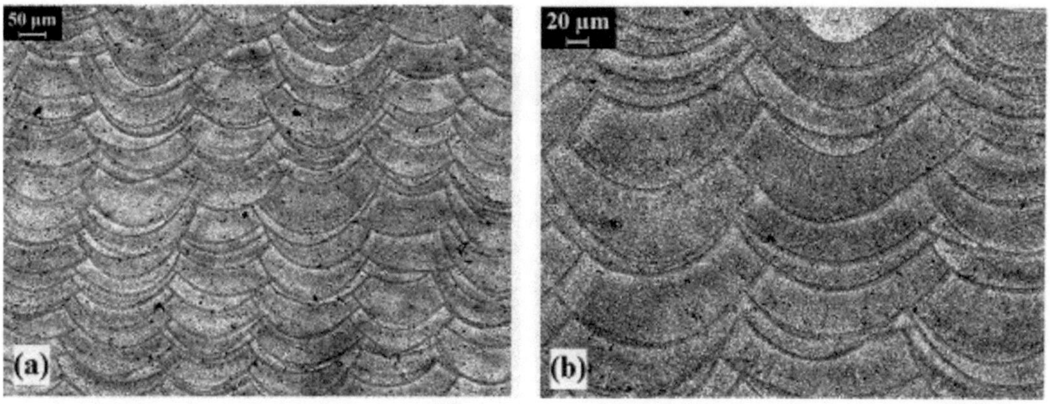

Figure 7. Optical micrographs of AlSi10Mg specimen produced by orienting scan tracks along x axis, after etching with Weck's reagent [51].

To appreciate the small grain size, the section along the build direction was observed by FESEM, focusing on a region between adjacent melt pools, as shown in Figure 8. After etching with HF and going at higher magnification, it can be seen that the melt pool contour region is characterized by a fine cellular-dendritic structure made by agglomerates of grains with mean diameters of a hundred of nanometers or less [37]. It was found that these agglomerates are different in length, thickness and aspect ratio, probably due to the different thermal heat fluxes during each scanning track.

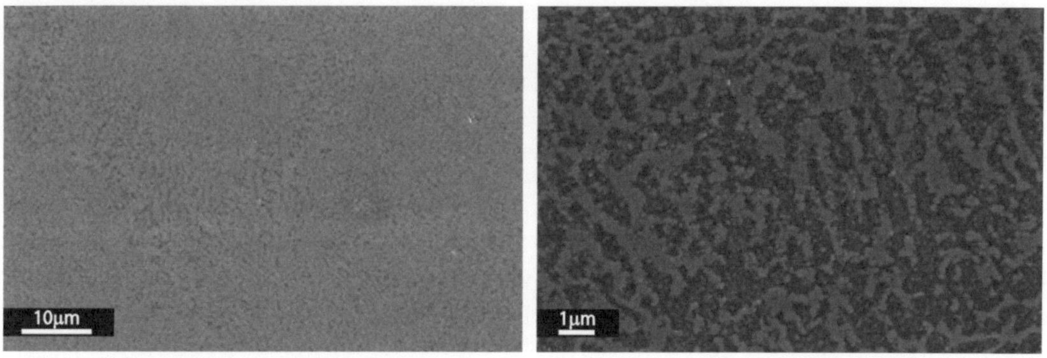

Figure 8. FESEM observations of AlSi10Mg specimen before etching (on the left), and after etching with HF (on the right) at higher magnification.

Also, fracture surfaces after tensile tests were investigated by FESEM, as reported in Figure 9. As can be seen, the surface is covered by very fine dimples, clearly visible only at high magnification. On the left image there can also be observed two little concave zones probably related to two spherical particles not completely melted. At higher magnification is possible to appreciate the very fine dimension of the microstructure, with presence of particles of tens of nanometers.

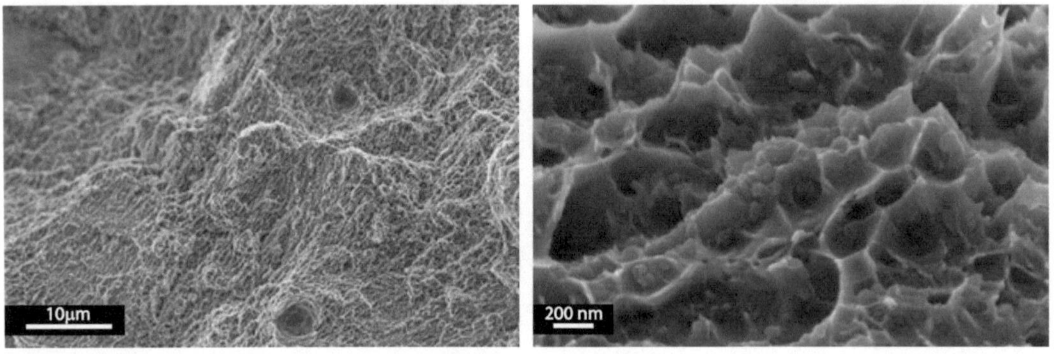

Figure 9. FESEM images of an aluminium alloy DMLS fracture surface covered by sub-micrometric voids and dimples with a nanometric size.

7.3. Effect of heat treatments

Due to the high thermal gradients, DMLS process fabricates parts with high residual thermal stresses. In order to avoid the bending of the AlSiMg parts fabricated, before removing them from the building platform, it is fundamental to perform an annealing for 2 h at 300 °C. Annealing offers very good dimensional and physical stability of the parts. It also provides the lowest level of residual stresses. The cooling is done in furnace or in the still air.

Apart from this, it is not common to perform a post heat treatment such as for conventional casting alloys. However, it is well known that mechanical properties of aluminium alloys are affected by heat treatments. Hence, the influence of heat treatments on hardness and tensile properties of AlSi10Mg parts produced by DMLS process was also investigated. In particular, annealing (T2), T4 and T6 heat treatments were carried out on fifteen AlSi10Mg samples for mechanical testing instead of stress relieving [51]. After the heat treatment, the hardness and tensile properties of these specimens were evaluated.

The first specimens were subjected to annealing treatment at 530 °C for 5 hours and cooled in the furnace. The second samples were obtained in T4 heat treatment: solution heat treated at the temperature of 530 °C for 5 hours and then quenched in water, followed by a room temperature ageing of at least two weeks. Finally the last samples were subjected to a T6 heat treatment cycle [49]. In this case, the AlSi10Mg specimens were solution heat treated at the temperature of 530 °C for 5 hours, then quenched in water and artificially aged at 160 °C for 12 hours.

The specimens were polished and the micro hardness was measured by using a Leitz instrument. The indentation load used was 50 g and duration of the indentation was 30 seconds. The results are illustrated in Table 8. The hardness after the T6 heat treatment is 13% higher than the mean value of the laser sintered AlSi10Mg samples just after the stress relieving treatment (105 HV, Table 6). Considering the T4 heat treatment, it reduced hardness by 15 % compared to AlSi10Mg sample without heat treatment, while with annealing the mean hardness was reduced by 58 %.

http://dx.doi.org/10.5772/58534

Type of heat treatment	Hardness (HV)	
	mean	S.D.
Annealing	46	3
T4	89	7
T6	119	6

Table 8. Effect of heat treatment on hardness of AlSi10Mg samples.

Tensile tests were also performed on AlSi10Mg after heat treatments. Again the specimens were produced according to ASTM E8M. The specimens were produced in the xy plane orientation by using DMLS process. Post processing operations such as shot peening or machining was not carried out on these specimens after heat treatments. The tensile tests of heat treated AlSi10Mg specimens were carried out by using EasyDur 3MZ-5000.

The results of the tensile tests are summarized in Table 9 [51].

Type of heat treatment	Rp0.2 (MPa)		UTS (MPa)		%A	
	mean	S.D	Mean	S.D	Mean	S.D
Annealing	72	7	113	3	12.6	0.9
T4	131	9	227	4	6.9	0.8
T6	245	8	278	2	3.6	0.8

Table 9. Tensile properties of AlSi10Mg specimens after heat treatments [51].

It can be concluded that all the heat treatments reduced the tensile properties of AlSi10Mg specimens (see also Table 7). Annealing reduced the ultimate tensile strength by 66 %, when comparing to the tensile strength of AlSi10Mg specimens in "as built" condition (just after DMLS, without shot peening). However, the elongation was improved by 103%, as expected. The tensile strength of the AlSi10Mg specimens after T4 treatment was reduced by 31% and ductility was improved by 11%, respectively. The ultimate tensile strength and elongation of the AlSi10Mg samples after T6 heat treatment were decreased by 16% and 42%, respectively. The only improvement observed after heat treatment is a slight improvement in yield strength, from 240 to 245 MPa, for the T6 samples with respect to the "as built" condition.

Stress-strain curves obtained from tensile tests of specimens after different heat treatment conditions are compared in Figure 10.

FESEM analysis of the fracture surfaces after different heat treatments were also performed, and the micrographs are shown in Figure 11. It can be seen that the dimensions of the dimples of fracture surface after each heat treatment considered are greater than the dimples of the corresponding fracture surface just after a stress relieving treatment. The total energy of the fracture is related to the size of the dimples [52]. Brandl et al. have reported that the SLM

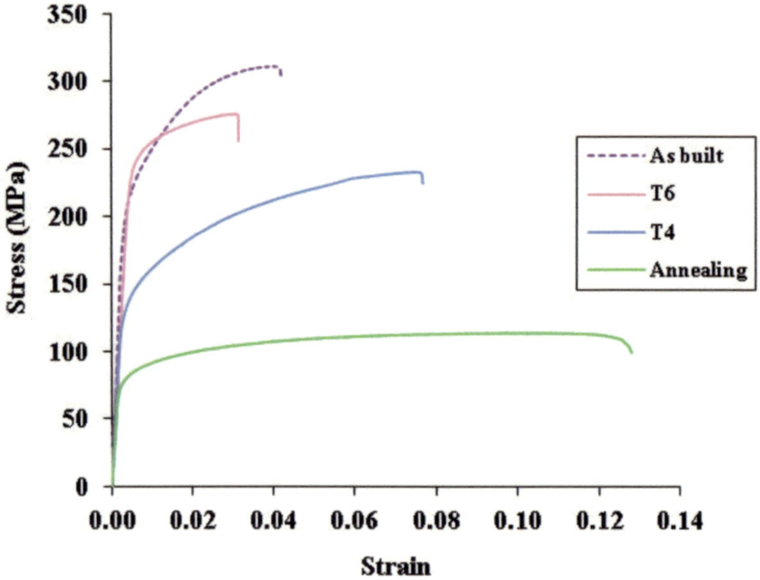

Figure 10. Stress vs strain curves of AlSi10Mg specimens in "as built" condition and after heat treatments [51].

processing of AlSi10Mg by maintaining the build platform at 300 °C and T6 heat treatment after SLM increase the fatigue resistance. This is due to homogenizing of microstructure after T6 heat treatment [53].

7.4. DMLS samples surface morphology

Due to its versatility of materials and shapes, the main advantage of DMLS is to produce metal complex-shaped components in one step, but it also has drawbacks that require careful process control: the high temperature gradients and densification ratio during the process yield high internal stresses or part distortion; the risk of balling and dross formation in the melt pool may result in bad surface roughness (from 8 to 20 μm without any post-treatment) [54-55]. The surface finish of a part, defined as the irregularities of any material resulting from machining operations, is critical in many applications, for example those requiring a surface roughness of 0.8 μm or better to avoid premature failure from surface initiated cracking [56]. There is no standard method to enhance the surface quality due to the complex nature of the process and to the different properties of the materials used [54,57]. Laser parameters like laser beam power, spot size and process parameters like scanning speed, hatching distance and layer thickness have a great influence on the quality of the laser sintered samples. The effect of some process parameters on AlSi10Mg parts produced by DMLS was studied through statistically designed experiments based on an L_{18} orthogonal array of Taguchi design [12]. Scan speed was found to have the greatest influence on the surface roughness. The Taguchi method uses *S/N* ratio to measure the variations of the experimental design. The equation of *smaller-the-better* was selected for the calculation of *S/N* ratio since it yields the lowest values of surface rough-

http://dx.doi.org/10.5772/58534

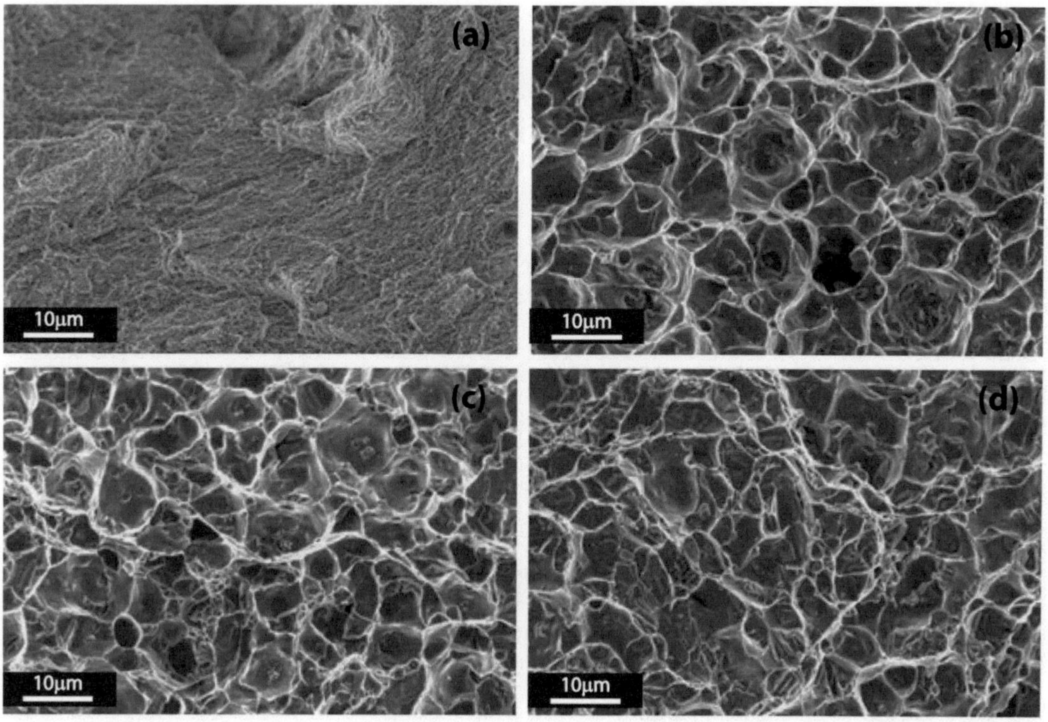

Figure 11. FESEM images at the same magnification of fracture surfaces after DMLS and a stress relieving treatment (a), after an annealing treatment – T2 (b), a T4 treatment (c) and a T6 treatment (d)

ness. It was found that *S/N* ratio is minimized when the scan speed is 900 mm/s, the laser power is 120 W, and the hatching distance is 0.10 mm.

The process of shot-peening involves a mechanical surface treatment whereby small balls impinge on the surface of the component. The repeated impacts not only induce compressive residual stress but also refine the microstructure at the surface and sub-surface region [58,59]. The near surface compressive residual stress field reduces the effective applied stresses of the component during application, which results in delayed crack initiation and retarded early crack propagation. It is important that shot-peening method is optimized to improve the depth of favorable compressive residual stress fields while minimizing surface roughening [59,60].

Figure 12 shows the surface roughness of a sample after the DMLS process (R_a=23.08 μm, R_z=152.92 μm) and then after the shot-peening post processing treatment (R_a=3.35 μm, R_z=31.81 μm). Thanks to the use of a 3D scanner is possible to analyze the entire surface (Figure 12a and 12d) and define the deviation of these from the best-fit 3D plane. The deviation of the sample before and after shot-peening is of 0.17 mm and 0.07 mm respectively (Figure 12b and 12e). Surface topography is classically characterized by surface profiles obtained via electronic contact profilometry (Figure 12 c and 12f).

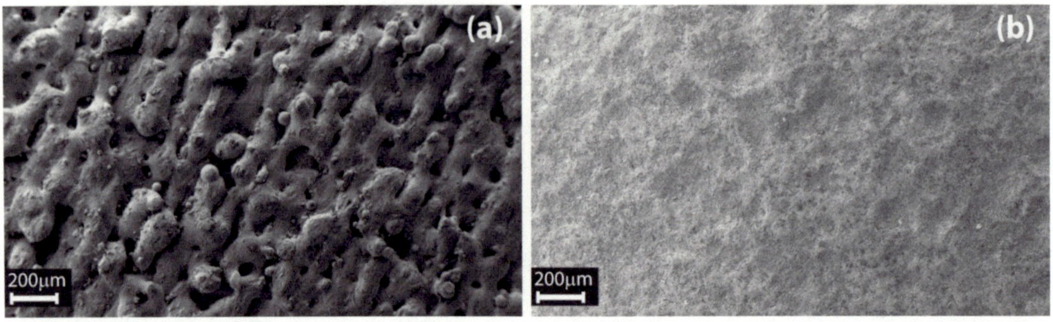

Figure 13. FESEM images of surface roughness of (a) sample with R_a=23 μm and (b) sample after shot peening at 8 bar.

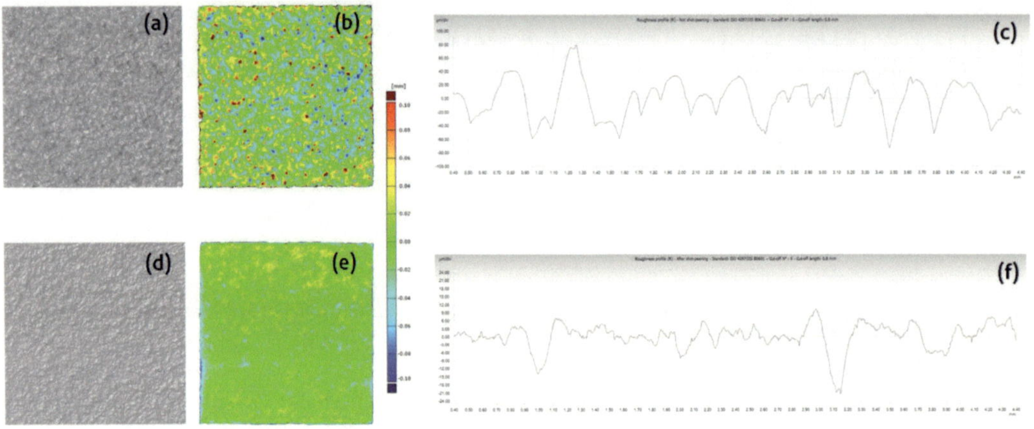

Figure 12. (a) 3D scan surface roughness of a DMLS AlSiMg sample before and (d) after shot-peening. (b) Deviation color maps respect to best fit plane before and (e) after shot-peening. (c) Roughness profile of the sample before and (f) after shot-peening.

Different roughness parameters are then extracted from the acquired surface profile. Among all the parameters for quantifying surface roughness based on tactile profile sections, R_a, the arithmetical mean deviation of the assessed profile, is by far the most extensive and most used parameter. It possible to see that shot peening with glass beads makes to significantly reduce the surface roughness. Different values of shot peening pressure were analyzed. The best results were obtained using a pressure of 8 bar: the surface roughness was reduced by up to 85%. These considerations on laser sintering parameters and the shot peening effect were confirmed by FESEM observations (Figure 13): the images give evidence of the improvement in surface roughness.

7.5. Lightweight metal structures by DMLS

One of the main interesting applications of additive manufacturing is for fabrication of customized, lightweight material structures, like periodic cellular lattice structures [61]. The word "cell" derives from the Latin word *cella* meaning a small compartment or enclosed space [62]. When nature constructs things, it is often done from many cells (cellula). Wood and cancellous bone are good examples of this: they are stiff, lightweight and multifunctional. From these natural inspirations a variety of synthetic cellular solids could be made including stochastic foams, consolidated powders, honeycombs and lattice blocks or miniature truss structures. These structures can offer high performance features such as high strength accompanied by a relatively low mass, good energy absorption characteristics and good thermal and acoustic insulation properties, making them suitable for high value aerospace, medical and engineering products [63]. The concept of designed cellular materials is motivated by the desire to put material only where it is needed for a specific application, as nature does. Leonardo da Vinci stated: "In her (nature's) inventions, nothing is lacking, and nothing is superfluous". From a mechanical engineering viewpoint, a key advantage offered by cellular materials is high strength accompanied by a relatively low mass.

In the past 15 years, the area of lattice materials has received considerable research attention due to their inherent advantages over foams in providing light, stiff, and strong materials. Many methods have been developed to analyze various cellular structures. Wang and McDowell have performed a comprehensive review of analytical modeling, mechanics, and characteristics of various metal honeycombs [64,65]. Deshpande et al. [66] have investigated extensively lattice cells, particularly the octet-truss structure.

However, the cellular lattice structures proposed and investigated in the previous studies could not exhibit good manufacturability in SLM. As reported in a recent study, the cellular structures with large unit cell sizes (greater than 5 mm) could not be built using the SLM process because overhanging struts in the cells led to the occurrence of serious deformation [67]. This adds considerable constraints on manufacturing versatile and complex cellular structures to meet requirements of different functions and applications, sacrificing the design freedom of cellular structures and geometrical capability of AM manufacturing. Sacrificial support structures might be used to support overhanging structures and thus prevent deformation. On the counter side, support structures inside complex cellular lattice structures are normally not acceptable because they are very difficult to remove, and waste materials and energy.

It was demonstrated, by the authors chapter, that it is possible to manufacture aluminium lattice structures by DMLS as the ones reported in Figure 14, with desired shape and internal features in a single fabrication step. This was possible because previously many experiments were focused on investigating the limitations of building surfaces without support structures. An overhanging structure is a part of a component that is not supported during building, by solidified material or a substrate on the bottom side. Consequently, the melt pool created by the heat input from the laser is supported by powder material. From this definition, it is clear that a part of a component is an overhanging structure depending on the orientation given to the part while building it.

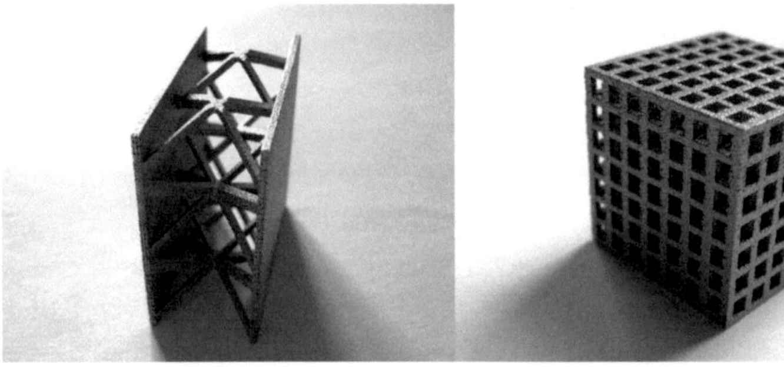

Figure 14. Lightweight components fabricated through DMLS in AlSiMg alloy.

In particular the cubic lattice structure of figure 14 was constructed with an angle of 45 degrees with respect to the building platform.

Following these design rules, a wide variety of different architectures can be made with fine control at the so called "cell level", as shown in Figure 15. The geometrical features selected for investigations were chosen because they could be the building blocks of more complex geometries. The structures have very good base metal properties thanks to the very fine microstructure typical of this process.

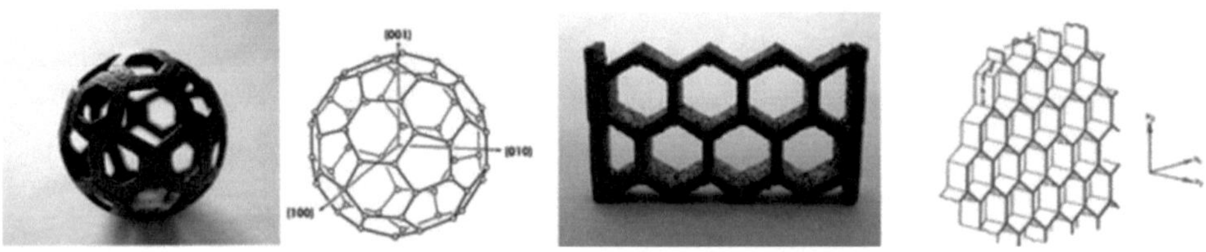

Figure 15. Cellular structures fabricated through DMLS in AlSiMg alloy: a fullerene structure (on the left), and a honeycomb cell structure (on the right).

8. Case study — Aluminium Matrix Composites (AMCs) by DMLS

As described in the previous paragraphs, aluminium alloys are currently produced by DMLS, in particular with compositions suitable for casting. The microstructure of these alloys is peculiar, since the very fast cooling occurring after the melting induced by laser provides an ultra-fine microstructure [34] and this effect is responsible for the significant increase of the mechanical properties that is observed with respect to conventional casting processes [37]. In fact, the conventional casting can be followed by a hardening heat treatment to improve its mechanical properties, treatment that seems not necessary in the case of DMLS process. Moreover, as a powder based process, DMLS also provides great opportunity to consolidate second or multiple material particles with metal powders to form novel metal matrix composites (MMCs). There is a growing research to develop MMCs via SLM process [68,69].

In order to be used inside a powder bed system, discontinuous ceramic particles are generally used. Among the most common reinforcement, silicon carbide (SiC) or aluminium oxide (Al_2O_3) are the most used, and in particular SiC-based AMCs are now widely used in many fields: from brake drums and cylinders liners of automobiles, to structural aerospace parts such as rotor vanes and plates. There are several methods to fabricate such composites: powder metallurgy, squeeze casting, stir casting or modified casting; however these conventional techniques are generally not suitable for complex shapes.

In the recent past different research groups investigated the feasibility to employ DMLS to obtain Al-based MMCs. Size and volume fraction of SiC particles have been varied to analyze the behavior of the composite [70], demonstrating that cracking occurs during the preparation by DMLS of these composites. In particular, crack density increases significantly after 15 volume percentage (vol.%) of SiC, and that there was not improvement of wear resistance over 20 vol.% of reinforcement. The same authors, in a previous study, found that microhardness increases with increase of volume fraction of SiC particles, as expected [39]. A couple of years before, Simchi et al. investigated the direct laser sintering of Al–7Si–0.3Mg/SiC composites [71], and showed that the densification rate increases at low SiC fractions but abruptly decreases at >~5 vol.%. Meanwhile, significant reaction occurs between the aluminium melt and the reinforcement particles, leading to formation of Al_4SiC_4 and silicon particles. However, these studies do not investigate the effects of process parameters on the density of metal matrix composites built by DMLS but the influence of design choices like supports generation and part orientation in the building chamber to avoid the presence of cracks during manufacture.

Starting from AlSiMg alloy, the possibility of composites fabrication by DMLS was investigated. The AlSiMg alloy has a high fluidity in the liquid phase, so that it is possible to prepare almost dense samples. Moreover, if silicon carbide is used a second phase, it is known in literature [72] that a high amount of silicon reduces or suppresses the reaction between aluminium and silicon carbide that brings to the formation of the dangerous aluminium carbide Al_4C_3, generally heavily detrimental for the mechanical properties. For this reason SiC particles were chosen as one of the possible reinforcing materials. α-SiC powders from H.C. Starck (UF-15, 15 m^2/g) were used, with density of 3.2 g/cm^3 and a mean particle size of 0.55 μm.

Another interesting reinforcement that was tested is the aluminium-magnesium spinel ($MgAl_2O_4$). This oxide has a high melting point of 2315 °C, a high Young's modulus and strength, and low thermal expansion coefficient (CTE). Moreover it is suggested in the literature [73] that it can improve substantially the creep behavior of composites if the size of the particles is very small. Thus nano-sized powders of $MgAl_2O_4$ were used (Nanocerox, mean size around 30 nm)

Since SiC particles are large while $MgAl_2O_4$ particles are much smaller, a different amount of the two reinforcements was used for the two composites, namely 10% and 1% in weight of SiC and $MgAl_2O_4$ respectively. To avoid deformation of the alloy spherical particles, simple mixing was done to prepare the composite powders, by using a ball milling system in ceramic jars, without any grinding medium, for 48 hours. The powders were then sieved with a mesh of 63 μm before putting them in the DMLS machine. The parameters used for the preparation of the composites are shown in Table 10.

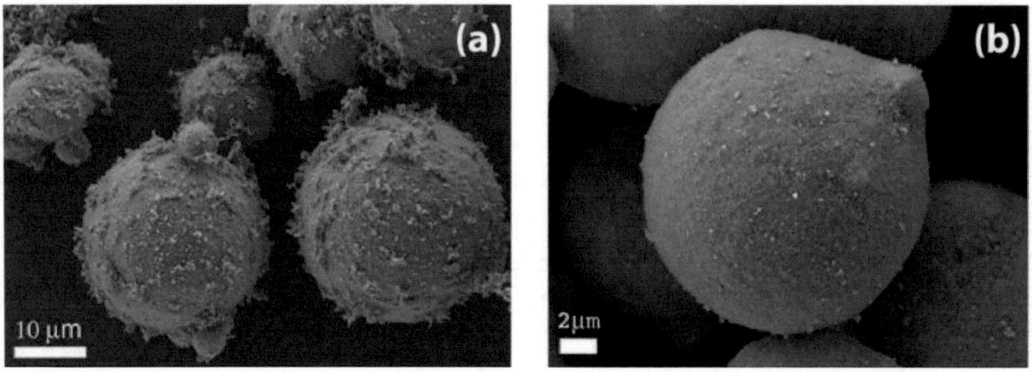

Figure 16. FESEM images of the composites powders: AlSiMg (a) and SiC (10%wt.) and (b) nano-$MgAl_2O_4$ (1%wt).

Parameters	AlSiMg	
	+ SiC	**+ $MgAl_2O_4$**
Scan speed (mm/s	500,700	600, 800
Laser Power (W)	180,195	195
Hatching dist. (mm)	0.17	0.17, 0.10

Table 10. Details of the process parameters employed.

In Figure 16 are shown the composite powders observed by FESEM. The metallic powders are not modified during the mixing process, however it is evident that both silicon carbide and nanospinel particles adhere on the surface of the aluminium alloy particles, and no agglomerates of pure ceramic particles are observed. In the case of silicon carbide the particles form a "fluffy" layer (Figure 16a), that causes an increased interaction between the surfaces of the powders. Thus, they have a different behavior in terms of flowability, causing some problems during the spreading of the powders, in particular in the first layers. This phenomenon is evident with silicon carbide, while it does not occur with the nanospinel (Figure 16b), probably due to the very small ceramic grain size and to the lower ceramic content. However it can be seen that the $MgAl_2O_4$ particles cover uniformly the round surface.

In the case of AlSiMg/SiC 10% composites, after a first step of parameter optimization, rectangular samples of 50x10x4 mm dimensions were obtained, with geometrical density between 2.59 and 2.61 g/cm^3. Since the theoretical density of these materials is 2.73 g/cm^3, the corresponding residual porosities are from 4.4 to 5%, a result in line with previous studies [39].

In the case of AlSiMg/SiC 10% composites, after a first step of parameter optimization, rectangular samples of 50x10x4 mm dimensions were obtained, with geometrical density between 2.59 and 2.61 g/cm^3. Since the theoretical density of these materials is 2.73 g/cm^3, the corresponding residual porosities are from 4.4 to 5%, a result in line with previous studies [39].
A problem however arises however with these composites: XRD measurements demonstrate

that silicon carbide is almost completely disappeared, while aluminium carbide is formed. This behavior is rather different to what happens for more conventional processes, where 10% Si is sufficient to suppress the Al-SiC reaction [72]. In this case, even if the contact time between molten alloy and ceramic is very low, the temperature is probably very high (in excess of 2000 K) and the reaction occurs. Ghosh et al. [39] already observed this phenomenon, but with a very low silicon content, while in the case of AlSiMg alloy the behavior was somewhat unexpected. The microstructure of the obtained composites is illustrated in the micrograph of Figure 17: it is very fine, as typical of the DMLS process, with elongated acicular grains, probably of Al_4C_3 as suggested by XRD analysis.

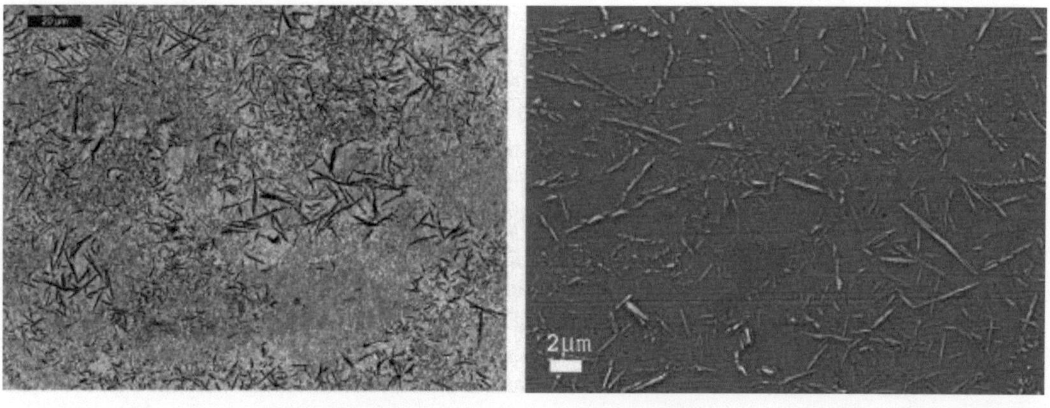

Figure 17. Optical (on the left) and FESEM (on the right) images of AlSiMg/SiC composite microstructure after polishing.

Residual SiC, instead, could be detected only in the very rare cases where some agglomerate is present, suggesting that the well dispersed single silicon carbide particle is instead completely reacted with the matrix. The presence of the ceramic reinforcement on hardness is in any case very evident: Brinell measurements mean value is 178 ± 2 HB for the SiC containing composites, which is 70% higher with respect to pure AlSiMg by DMLS.

In the case of nano-$MgAl_2O_4$ AMCs, the density is more depending on the specific choice of parameters than in the case of silicon carbide containing composites. In this case small samples were obtained to perform a more detailed screening of the parameters, and for this reason samples with 15x15x10 mm size were fabricated. With the parameters optimized in the best range, the measured average densities of the samples varied between 2.59 and 2.63 g/cm^3, corresponding to residual porosities in the range from 2.2 to 3.5% (the theoretical density is 2.689 g/cm^3).

In this case, the XRD analysis shows the presence of Al, Si, with some extremely small peaks that were attributed to the inter-metallic phase Mg_2Si and to some mixed metal-oxide of unknown origin ($AlSiO_x$), with a spectrum that is rather similar to the one of the pure alloy. Indeed, in this case the XRD technique is clearly less powerful to recognize possible reactions, due to the low content of reinforcing ceramic. The microstructure of these composites is also very similar to the case of the pure aluminium matrix, as illustrated in the micrographs of

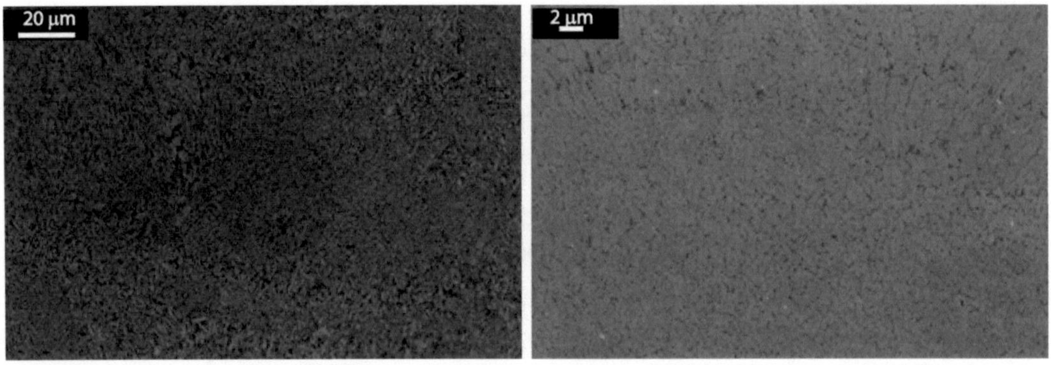

Figure 18. Optical (on the left) and FESEM (on the right) images of AlSiMg/Mg-nanospinel composite microstructures.

Figure 18. As typical of the DMLS process the microstructure is very fine, made by submicrometric grains, elongated in the correspondence of the melt pools contour due to the heat transfer flux direction. A careful observation however shows a higher inhomogeneity of these composites with respect to the pure aluminium alloy, suggesting that the presence of nano-$MgAl_2O_4$ particles changes the behavior during solidification. This inhomogeneity, together with the higher residual porosity, reflects on the values of Brinell Hardness, that is 93 ± 3 HB, 11% lower than the pure AlSiMg alloy processed through DMLS.

Acknowledgements

The authors wish to acknowledge financial support from the AMAZE Project, which is co-funded by the European Commission in the 7th Framework Programme (contract FP7-2012-NMP-ICT-FoF-313781), by the European Space Agency and by the individual partner organizations.

Author details

Diego Manfredi[1], Flaviana Calignano[1], Manickavasagam Krishnan[1,3], Riccardo Canali[1,2], Elisa Paola Ambrosio[1], Sara Biamino[2], Daniele Ugues[2], Matteo Pavese[2] and Paolo Fino[1,2]

1 Center for Space Human Robotics @Polito, Istituto Italiano di Tecnologia, Corso Trento, Torino, Italy

2 DISAT-Dipartimento Scienza Applicata e Tecnologia, Politecnico di Torino, Corso Duca degli Abruzzi, Torino, Italy

3 DIGEP – Dipartimento di Ingegneria Gestionale e della Produzione, Politecnico di Torino, Corso Duca degli Abruzzi, Torino, Italy

References

[1] Gibson I, Rosen WD, Stucker B. Additive Manufacturing Technologies-Rapid Prototyping to Direct Digital Manufacturing. Springer; 2010.

[2] Wohlers T. Wohlers Report 2011 State of the Industry, Annual Worldwide Progress Report. Wohlers Associates; 2011.

[3] Campbell I, Bourell D, Gibson I. Additive manufacturing: Rapid prototyping comes of age. Rapid Prototyping Journal 2012; 18 (4): 255–258.

[4] http://www.amaze-project.eu/

[5] http://www.rapolac.eu/

[6] Gerrits A, Jones CL & Valero R. Custom-Fit: Quality of Life of European Sporting Public through Custom-Fit Products. The Engineering of Sport. 2006, 6: 5-10.

[7] http://www.e-break.eu/

[8] http://www.tialcharger.de/

[9] Wohlers T. Additive Manufacturing Advances. Manuf. Eng. 2012; 148 (4): 55–63

[10] Hot-Fire Tests Show 3-D Printed Rocket Parts Rival Traditionally Manufactured Parts. 2013. [Online]. Available: http://www.nasa.gov/exploration/systems/sls/3dprinting.html [Accessed: 13-Aug-2013].

[11] ASTM F2792-12a: Standard Terminology for Additive Manufacturing Technologies. DOI:10.1520/F2792-12A

[12] Calignano F, Manfredi D, Ambrosio EP, Iuliano L, Fino P. Influence of process parameters on surface roughness of aluminium parts produced by DMLS. Int. J. Adv. Manuf. Technol. 2013; 67 (9–12): 2743–2751.

[13] Aumund-Kopp C, Petzoldt F. Laser Sintering of parts with complex internal structures. In: Proceedings of the 2008 world congress on powder metallurgy & particulate materials. 2008; 1 (3): 385–397.

[14] Tsopanos S, Mines RAV, McKown S, Shen Y, Cantwell WJ, Brooks W, Sutcliffe CJ. The Influence of Processing Parameters on the Mechanical Properties of Selectively Laser Melted Stainless Steel Microlattice Structures. J. Manuf. Sci. Eng. 2010; 132 (4): 041011.

[15] Childs THC, Hauser C, Badrossamay M. Selective laser sintering (melting) of stainless and tool steel powders: experiments and modelling. Proc. Inst. Mech. Eng. Part B J. Eng. Manuf. 2005; 219 (4): 339–357.

[16] Tolosa I, Garciandía F, Zubiri F, Zapirain F, Esnaola A. Study of mechanical properties of AISI 316 stainless steel processed by 'selective laser melting', following different manufacturing strategies. Int. J. Adv. Manuf. Technol. 2010; 51 (5–8): 639–647.

[17] Li R, Liu J, Shi Y, Du M, Xie Z. 316L stainless steel with gradient porosity fabricated by selective laser melting. J. Mater. Eng. Perform. 2010; 19 (5): 666–671.

[18] Averyanova M, Cicala E, Bertrand P, Grevey D. Experimental design approach to optimize selective laser melting of martensitic 17-4 PH powder: part I – single laser tracks and first layer. Rapid Prototyping Journal. 2012; 18 (1): 28–37.

[19] Dewidar MM, Dalgarno KW, Wright CS. Processing conditions and mechanical properties of high-speed steel parts fabricated using direct selective laser sintering. Proc. Inst. Mech. Eng. Part B J. Eng. Manuf. 2003; 217 (12): 1651–1663.

[20] Yasa E, Deckers J, Kruth JP, Rombouts M, Luyten J. Charpy impact testing of metallic selective laser melting parts. Virtual Phys. Prototyp. 2010; 5 (2): 89–98.

[21] Li X, Wang C, Zhang W, Li Y. Fabrication and compressive properties of Ti6Al4V implant with honeycomb-like structure for biomedical applications. Rapid Prototyping Journal. 2010; 16 (1): 44–49.

[22] Vrancken B, Thijs L, Kruth JP, Van Humbeeck J. Heat treatment of Ti6Al4V produced by Selective Laser Melting: Microstructure and mechanical properties. J. Alloys Compd. 2012; 541: 177–185.

[23] Van Hooreweder B, Moens D, Boonen R, Kruth JP, Sas P. Analysis of Fracture Toughness and Crack Propagation of Ti6Al4V Produced by Selective Laser Melting. Adv. Eng. Mater. 2012; 14 (1–2): 92–97.

[24] Song B, Dong S, Zhang B, Liao H, Coddet C. Effects of processing parameters on microstructure and mechanical property of selective laser melted Ti6Al4V. Mater. Des. 2012; 35: 120–125.

[25] Marcu T, Todea M, Gligor I, Berce P, Popa C. Effect of surface conditioning on the flowability of Ti6Al7Nb powder for selective laser melting applications. Appl. Surf. Sci. 2012; 258 (7): 3276–3282.

[26] Wang Z, Guan K, Gao M, Li X, Chen X, Zeng X. The microstructure and mechanical properties of deposited-IN718 by selective laser melting. J. Alloys Compd. 2012; 513: 518–523.

[27] Yadroitsev I, Pavlov M, Bertrand P, Smurov I. Mechanical properties of samples fabricated by selective laser melting. In: 14èmes Assises Européennes du Prototypage & Fabrication Rapide, 2009.

[28] Kumar S, Pityana S. Laser-based additive manufacturing of metals. Adv. Mater. Res. 2011; 227: 92–95.

[29] Kumar S. Selective Laser Sintering: Recent Advances. In: 4th Pacific International Conference on Applications of Lasers and Optics. Wuhan-China, 2010.

[30] Khan M, Dickens PM. Processing parameters for Selective Laser Melting (SLM) of gold. In: Proceedings of Solid Freeform Fabrication symposium, 2008: 278–289.

[31] Louvis E, Fox P, Sutcliffe CJ. Selective laser melting of aluminium components. J. Mater. Process. Technol. 2011; 211 (2): 275–284.

[32] Olakanmi EO, Dalgarno KW, Cochrane RF. Laser sintering of blended Al-Si powders. Rapid Prototyping Journal. 2012; 18 (2): 109–119.

[33] Olakanmi EO, Cochrane RF, Dalgarno KW. Densification mechanism and microstructural evolution in selective laser sintering of Al – 12Si powders. J. Mater. Process. Technol. 2011; 211 (1): 113–121.

[34] Thijs L, Kempen K, Kruth JP, Van Humbeeck J. Fine-structured aluminium products with controllable texture by selective laser melting of pre-alloyed AlSi10Mg powder. Acta Mater. 2013; 61 (5): 1809–1819.

[35] Kempen K, Thijs L, Yasa E, Badrosamay M, Verheecke W, Kruth JP. Process optimization and microstructural analysis for selective laser melting of AlSi10Mg. In: Proceedings of Solid Freeform Fabrication symposium, 2011: 484–495.

[36] Manfredi D, Calignano F, Ambrosio EP, Krishnan M, Canali R, Biamino S, Pavese M, Atzeni E, Iuliano L, Fino P and Badini C. Direct Metal Laser Sintering: an additive manufacturing technology ready to produce lightweight structural parts for robotic applications. La Metallurgia Italiana. 2013; 105 (10): 15-24.

[37] Manfredi D, Calignano F, Krishnan M, Canali R, Ambrosio EP, Atzeni E. From Powders to Dense Metal Parts: Characterization of a Commercial AlSiMg Alloy Processed through Direct Metal Laser Sintering. Materials. 2013; 6 (3): 856-869.

[38] Kumar S, Kruth JP. Composites by rapid prototyping technology. Mater. Des. 2010; 31 (2): 850–856.

[39] Ghosh SK, Saha P, Kishore S. Influence of size and volume fraction of SiC particulates on properties of ex situ reinforced Al-4.5Cu-3Mg metal matrix composite prepared by direct metal laser sintering process. Mater. Sci. Eng. A, 2010; 527 (18–19): 4694–4701.

[40] Thijs L, Verhaeghe F, Craeghs T, Humbeeck JV, Kruth JP. A study of the microstructural evolution during selective laser melting of Ti-6Al-4V. Acta Mater. 2010; 58 (9): 3303–3312.

[41] Spierings AB, Herres N, Levy G. Influence of the particle size distribution on surface quality and mechanical properties in AM steel parts. Rapid Prototyping Journal. 2011; 17 (3): 195–202.

[42] Buchbinder D, Schleifenbaum H, Heidrich S, Meiners W, Bültmann J. High Power Selective Laser Melting (HP SLM) of Aluminum Parts. Phys. Procedia 2011; 12: 271–278.

[43] Qiu C, Adkins NJE, Attallah MM. Microstructure and tensile properties of selectively laser-melted and of HIPed laser-melted Ti-6Al-4V. Mater. Sci. Eng. A, 2013; 578: 230–239.

[44] Thijs L, Montero Sistiaga ML, Wauthle R, Xie Q, Kruth JP, Van Humbeeck J. Strong morphological and crystallographic texture and resulting yield strength anisotropy in selective laser melted tantalum. Acta Mater., 2013; 61 (12): 4657–4668.

[45] Kruth JP, Badrosamay M, Yasa E, Deckers J, Thijs L, Van Humbeeck J. Part and material properties in selective laser melting of metals. In: Proceedings of the 16th International Symposium on Electromachining (ISEM XVI), 2010.

[46] Shiomil M, Yamashital T, Osakada K, Shiomi M, Yamashita T, Abe F, Nakamura K. Residual Stress within Metallic Model Made by Selective Laser Melting Process. Ann. CIRP, 2014; 53 (1): 195–198.

[47] Zaeh MF, Branner G. Investigations on residual stresses and deformations in selective laser melting. Prod. Eng., 2009; 4 (1): 35–45.

[48] Kruth JP, Yasa E, Deckers J. Roughness improvement in Selective Laser Melting. In: Proceedings of the 3rd International Conference on Polymers and Moulds Innovations. PMI, 2008: 170–183.

[49] ASM Handbook, Volume 2-Properties and Selection: Non ferrous Alloys and Special-Purpose Materials, and Volume 4 – Heat treating. 1990, ASM International The Materials Information Company, United States of America, ISBN 0-87170-379-3.

[50] Hauser C, Childs THC, Badrossamay M. Further developments in process mapping and modelling in direct metal selective laser melting. In: Proceedings of the Solid Freeform Fabrication symposium, 2004; 448–459

[51] Krishnan M. Investigation of material and mechanical properties of Al alloy and Al based MMC parts produced by DMLS for industrial application. PhD Thesis. Politecnico di Torino; 2014.

[52] Hertzberg RW. Deformation and Fracture Mechanics of Engineering Materials. John Wiley & Sons Inc, 1996.

[53] Brandl E, Heckenberger U, Holzinger V, Buchbinder D. Additive manufactured AlSi10Mg samples using Selective Laser Melting (SLM) Microstructure, high cycle fatigue, and fracture behavior. Materials and Design. 2012; 34: 159-169.

[54] Delgado J, Ciurana J, Rodríguez CA. Influence of process parameters on part quality and mechanical properties for DMLS and SLM with iron-based materials. Int J Adv Manuf Technol. 2012; 60:601–610

[55] Kruth JP, Levy G, Klocke F, Childs THC. Consolidation phenomena in laser and powder-bed based layered manufacturing. CIRP 2007; 56(2):730–759

[56] Dalgarno K. Materials research to support high performance RM parts. Rapid Manufacturing 2nd International Conference, Loughborough University, 2007; 147–56

[57] Sachdeva A, Singh S, Sharma VS. Investigating surface roughness of parts produced by SLS process. Int J Adv Manuf Technol. 2012.

[58] Evans A, Bruno G. Relaxation of residual stress in shot peened Udimet 720Li under high temperature isothermal fatigue. Int J Fatigue. 2005; 27: 1530–1534.

[59] Turski M, Clitheroe S, Evans AD, Rodopoulos C. Engineering the residual stress state and microstructure of stainless steel with mechanical surface treatments. Appl Phys A. 2010; 99: 549–56.

[60] Murr LE, Gaytan SM, Ramirez DA, Martinez E, Hernandez J, Amato KN, Shindo PW, Medina FR, Wicker RB. Metal Fabrication by Additive Manufacturing Using Laser and Electron Beam Melting Technologies. J. Mater. Science Technology. 2012; 28(1): 1–14.

[61] Ashby MF et al., Metals Foams: A Design Guide. Woburn, MA: Butterworth-Heinemann, 2000.

[62] Gibson LJ, Ashby MF. Cellular Solids, Structure and Properties. 2nd Edition, Cambridge University Press, Cambridge, 1997.

[63] Williams CB, Cochran JK, Rosen DW. Additive manufacturing of metallic cellular materials via three-dimensional printing. Int. J. Adv. Manuf. Technol. 2011; 53: 231–239.

[64] Wang AJ, McDowell DL, Optimization of a metal honeycomb sandwich beam-bar subjected to torsion and bending. Int. J. Solids Struct. 2002; 40(9): 2085-2099.

[65] Wang AJ, McDowell DL. Yield surfaces of various periodic metal honeycombs at intermediate relative density. Int. J. Plasticity. 2005; 21(2): 285-320.

[66] Deshpande VS, Fleck NA, Ashby MF. Effective properties of the octet-truss lattice material. Journal of Mechanics and Physics of Solids. 2001; 49(8): 1747-1769.

[67] Yan C, Hao L, Hussein A, Raymont D. Evaluations of cellular lattice structures manufactured using selective laser melting. International Journal of Machine Tools & Manufacture. 2012; 62: 32–38.

[68] Kumar S, Kruth JP. Composites by rapid prototyping technology. Materials and Design. 2010; 31: 850-856.

[69] Hao L, Dadbakhsh S. Materials and process aspects of selective laser melting of metals and metal matrix composites: a review. Chinese Journal of Lasers. 2009; 36 (12): 3192–3203.

[70] Ghosh SK, Saha P. Crack and wear behavior of SiC particulate reinforced aluminium based metal matrix composite fabricated by direct metal laser sintering process. Materials and Design. 2011; 32: 139-145.

[71] Simchi A, Godlinski D. Effect of SiC particles on the laser sintering of Al–7Si–0.3Mg alloy. Scripta Materialia. 2008; 59: 199-202.

[72] Narciso J, García-Cordovilla C, Louis E. Reactivity of thermally oxidized and unoxidized SiC particulates with aluminium-silicon alloys. Materials Science and Engineering B. 1992; 15: 148-155

[73] De Vita A, Roitti S, Sbaizero O. Creep of aluminium metal matrix composites reinforced with magnesium spinel. Key Engineering Materials. 2004; 264-268: 953-956.

Chapter 2

Mechanical Behavior of Precipitation Hardened Aluminum Alloys Welds

R.R. Ambriz and D. Jaramillo

Additional information is available at the end of the chapter

http://dx.doi.org/10.5772/58418

1. Introduction

Due to its relative low density and high strength, the 2*xxx*, 6*xxx* and 7*xxx* aluminum alloys series are largely used in transportation industry to produce structural frames and components. These alloys find application in artificial aging condition which allows obtaining high mechanical properties, i.e. yield strength of 7075-T651 aluminum alloy as high as 500 MPa. However, when welding processes are used to perform a junction, a large amount of the heat input is dissipated, by heat conduction, throughout the base material close to the welding bead. This thermal dissipation induces localized isothermal sections where the thermal gradient have an important and detrimental effect on the microstructure and therefore on the mechanical properties of the welded joint (soft zone formation). This microstructural change affects the performance in service of the welded joints, since mechanical properties reduce drastically with respect to base material.

This chapter provides information about mechanical behavior of welded joints of aluminum alloys in terms of properties, determined by tensile, indention and fatigue tests, as well as, the fatigue crack growth conditions in different zones of the welded joints.

1.1. Precipitation and mechanical properties of aluminum alloys

The precipitation hardening process requires that the second component in the aluminum alloy, is sufficiently soluble to allow extensive dissolution at an elevated temperature (solubilization treatment temperature) and that the solubility is considerably reduced at lower temperatures, such is the case of Al-Cu alloys (Figure 1) [1].

According to Figure 1, to induce precipitation hardening, the alloy is heated at a temperature higher than the solvus temperature to produce a homogeneous solid solution α, allowing dissolution of the second phase θ and eliminates the segregation of the alloy. Once, the solubilization temperature is reached, the alloy is cooling at high rate in order to limit the diffusion process of the atoms toward potential sites of nucleation. Finally, the supersaturated solid solution α_{ss} is heated at a temperature below the solvus temperature. At this temperature, the atoms have the ability to diffuse at short distances. Because the α_{ss} phase is not stable, the atoms of Cu diffuse at several sites of nucleation and a control precipitation could be formed. Precipitation hardening in metals is performed to produce a particulate dispersion of second phases to generate obstacles for the dislocation movement. The degree of hardening depends of the metallic system, the volume fraction and size of the particles and the interaction of the particles with dislocations. The interactions of the precipitated particles with dislocations are very important in terms of the magnitude of hardening. Some mechanisms have been established, involving particle bypassing by Orowan looping, bypass slip, or particle shearing. Figure 2, shows the forces acting on a mobile dislocation in a stressed metal containing a dispersion of second phase particles.

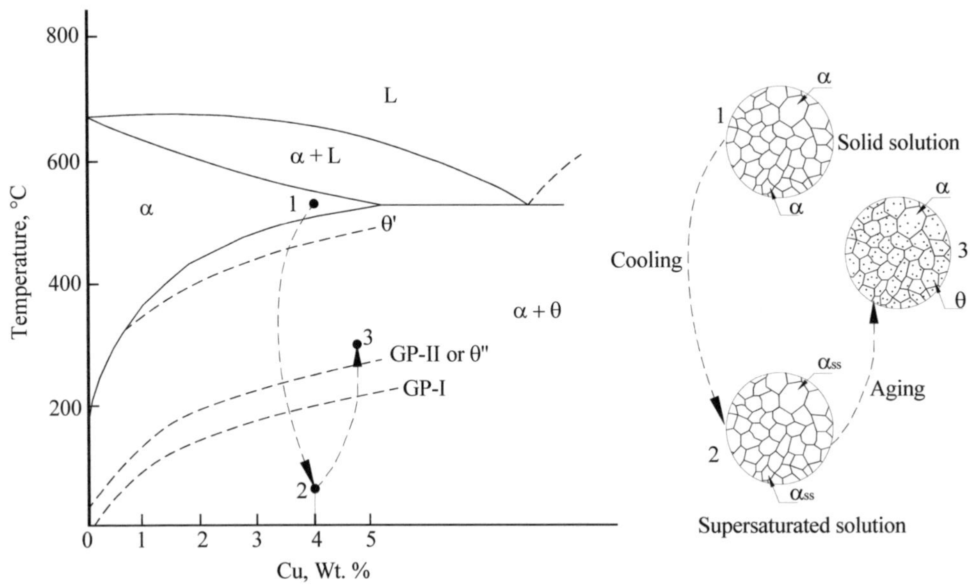

Figure 1. Aluminum-cooper phase diagram rich in aluminum showing the solubilization and precipitation process [1].

Considering the equilibrium forces between the line tension T of the dislocation and the resistance force of the second phase particle F, it is obtained:

$$F = 2T \sin \theta \tag{1}$$

As F increases, so the bowing of the dislocations increases, i.e. θ increases. The magnitude of F is important in controlling the sequence of events. The dislocation line tension force is

http://dx.doi.org/10.5772/58418

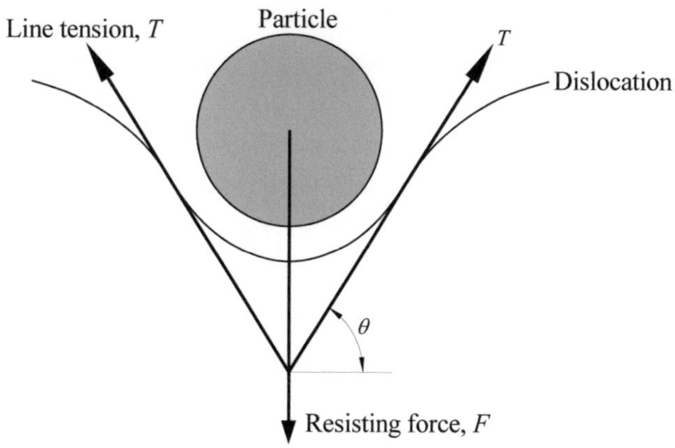

Figure 2. Balance of forces acting during particle resistance to dislocation movement [1].

maximal when θ=90º. If the particle is hard, such that F can be greater than $2T$, then dislocations will bypass the particle either by Orowan looping or cross-slip and the particle will remain unchanged, i.e. non deformed (Figure 3).

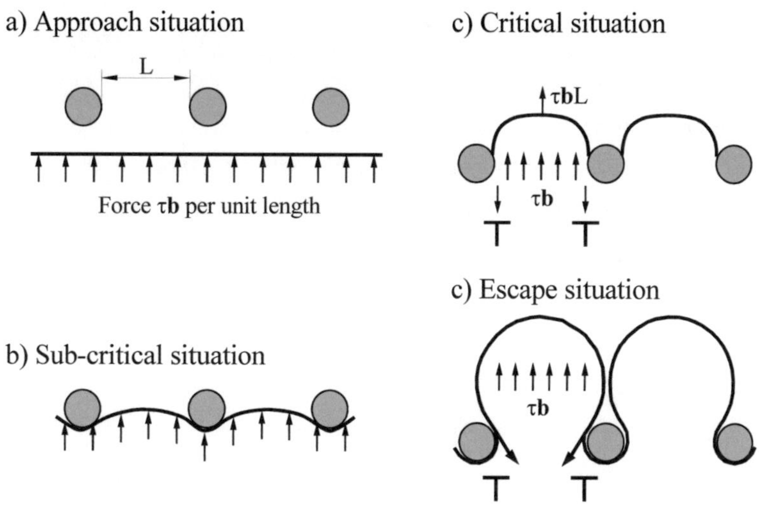

Figure 3. Dislocation meets hard undeformable second phase particles: dislocation release at higher stresses may occur by Orowan looping or by cross-slip [1].

The actual strength of the particle under this circumstance becomes irrelevant, as the bypassing operation becomes dependent only upon the interparticle spacing. If, however, the strength of the particle is such that the maximum resistance force is attained before sin θ=1, then particles will be sheared and the dislocation will pass through the particle (Figure 4).

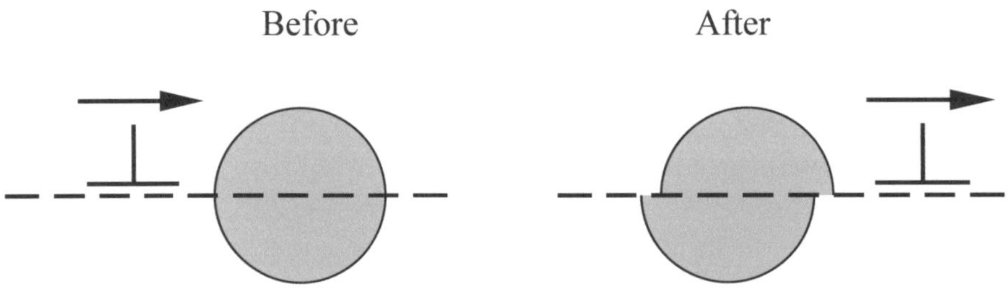

Figure 4. Dislocation motion may continue through second phase particles (particle cutting).

Therefore, it follows that, for a given interparticle spacing (given volume fraction and particle size), hard particles will give the maximum precipitation hardening, and this condition defines the maximum degree of hardening attainable. Soft particles give a lesser degree of hardening. Consideration of the relationship between the applied stress and the dislocation bowing, following Orowan [2], leads to the Orowan equation:

$$\Delta\tau_y = \frac{G\mathbf{b}}{L} \tag{2}$$

where $\Delta\tau_y$ is the increase in yield stress due to the particles, G the shear modulus of the matrix, **b** the Burgers vector of the dislocation, and L the particle spacing. The L in the Orowan equation is usually considered to be the distance between particles arranged on a square grid in the slip plane.

Ashby [3] further developed his equation to take into account the interparticle spacing, and the effects of statistically distributed particles. The Ashby-Orowan relationship is given as:

$$\Delta\tau_y = 0.84\left(\frac{1.2G\mathbf{b}}{2\pi L}\right)\ln\left(\frac{x}{2\mathbf{b}}\right) \tag{3}$$

Application of the Taylor factor for polycrystalline materials, expressing the microstructural parameters in terms of the volume fraction and real diameter and converting shear stress to tensile stress, yields [4]:

$$\Delta\sigma_y = \left(\frac{0.538G\mathbf{b}f^{1/2}}{X}\right)\ln\left(\frac{X}{2\mathbf{b}}\right) \tag{4}$$

where $\Delta\sigma_y$ is the increase in yield strength (MPa), G is the shear modulus (MPa), **b** is the Burgers vector (mm), f is the volume fraction of particles, and X is the real (spatial) diameter of the particles (mm). Table I presents the hardening effect in terms of yield strength for some aluminum alloys in annealed (O) and artificial age hardened conditions (T6).

	Yield strength (MPa)		
Alloy	*Annealed condition (O)*	*Artificial age hardened condition (T6)*	*Difference*
2014	97	417	320
6061	55	276	221
7075	103	503	400

Table 1. Hardening increments in aluminum alloys [5].

1.2. Welding in age hardened condition

The *2xxx*, *6xxx* and *7xxx* aluminum alloys are known to have a strong tendency to overage during welding, especially in fully aged condition (T6). A schematic representation of the microstructural changes in welding of aluminum alloys is shown in Figure 5 [6]. During welding the adjacent metal to the fusion zone (welding bead) is heated and the heat affected zone (HAZ) contents two principal zones. The zone of lower temperature near to the base metal is exposed to a range of temperatures where the aging phenomena and overaging may occur. The zone of higher temperature is treated by solubilization assuming high cooling rate, the effects are less severe, because the microstructure of the material will tend to age in natural manner. However, and isothermal zone appears at within the HAZ for which the obtained temperature is located between two well defined temperatures, i.e. the artificial aging temperature of the alloy and the solvus temperature. As a result a microstructural transformation takes places due to the thermodynamic instability of the precipitates. For instance, in the case of the 6061-T6 alloy, the over-aging in the HAZ is produced by the transformation of the β″ (needle shape) precipitates into β′ (rod shape) precipitates according to the following precipitation sequence [7]:

$$\alpha_{ss} \rightarrow \text{solute clusters} \rightarrow \text{GP zones} \rightarrow \beta'' \text{ (needle)} \rightarrow \beta' \text{ (rod)} \rightarrow \beta$$

Figure 6, shows the weld thermal cycles for gas metal arc welding (GMAW) process in a 6061-T6 (Al-Si-Mg) alloy at different preheat conditions and their correlation between the C precipitation curve. In this sense, Myhr et al. [9] studied the microstructural evolution during the cooling weld thermal cycle in Al-Si-Mg alloys (Figure 7). They determined that when the peak temperature T_p approaches 315 °C during a period of time of 10 s, the microstructure consists of a mixture of coarse rod-shape β′ and fine needle-shape β″ precipitates as shown in Figure 7b and c. The transformation from β″ to β′ precipitates occurs to an increasing extent with increasing peak temperatures. At T_p of 390 °C the β′ is the dominant microstructural constituent, as indicated in Figure 7d.

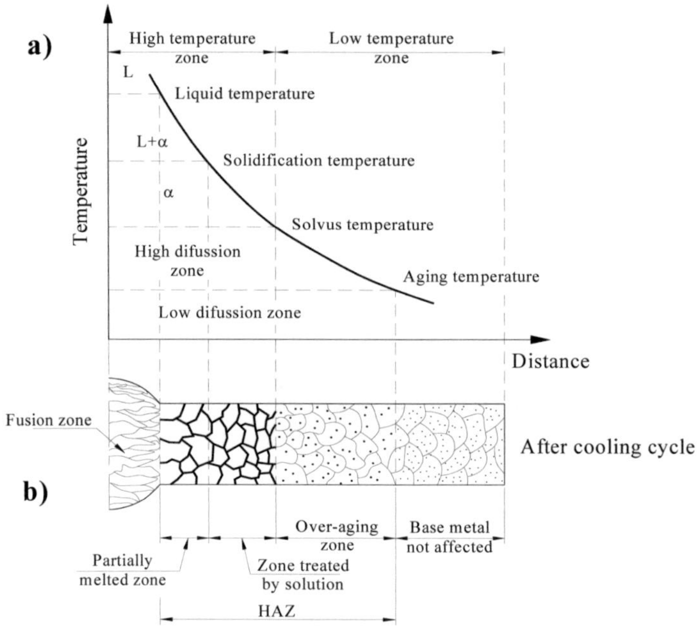

Figure 5. Schematic representation of the microstructural changes in heat treatable aluminum alloys during a fusion welding process, a) cooling thermal cycle from peak temperature, and b) microstructure of the welded joint at ambient temperature [8].

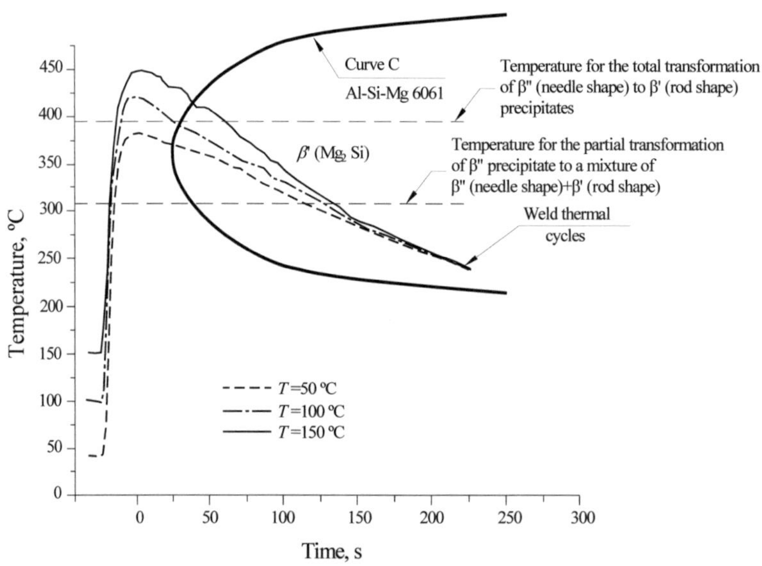

Figure 6. Correlation between weld thermal cycles in gas metal arc welds with three different preheating conditions, and the C precipitation curve for a 6061-T6 aluminum alloy [8].

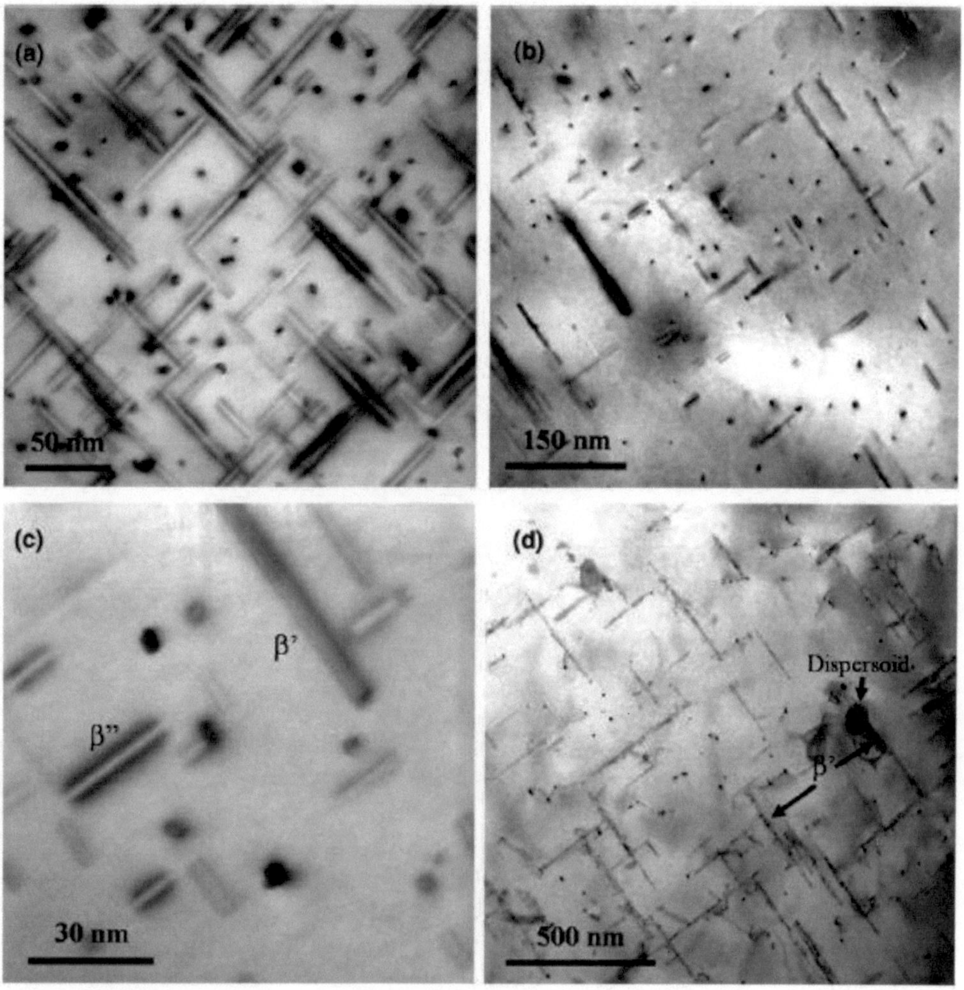

Figure 7. TEM bright field images of microstructures observed in the <100> Al zone axis orientation after artificial aging and Gleeble simulation (Series 1), (a) needle-shaped β'' precipitates which form after artificial aging, (b) mixture of coarse rod-shaped β′particles and fine needle-shaped β′′ precipitates which form after subsequent thermal cycling to T_p=315 °C (10 s holding time), (c) close up of the same precipitates shown in (b) above, and (d) coarse rod-shaped β' particles which form after thermal cycling to T_p=390 °C (10 s holding time) [9].

In addition to the microstructural transformation, after welding porosity and liquation cracking could also exist, which affect directly the mechanical behavior of the welded joints.

Porosity in welds of aluminum alloys is very complicated to control, because of the high hydrogen diffusion in liquid aluminum (Figure 8), the environmental interaction and the high rate of solidification.

When a fusion welding process in aluminum is performed, the hydrogen diffusion in melted metal could be as high as 1.00 cm^3/g, this fact produces the formation of gas bubbles. If we consider the bubble formation in liquid metal as schematically represented in Figure 9, the

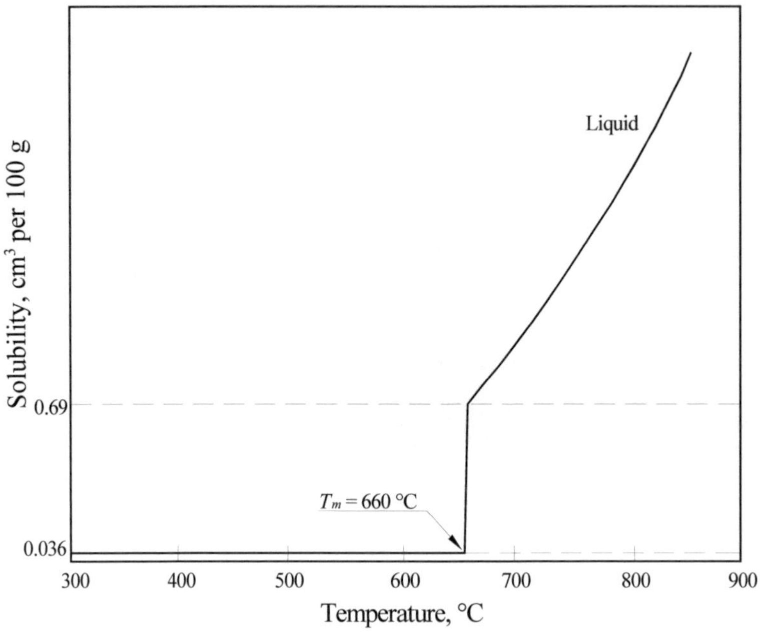

Figure 8. Hydrogen solubility in aluminum [10].

bubble begins to ascend when the surface tension is overcome by the buoyant force, which tends to push the bubble to the free surface.

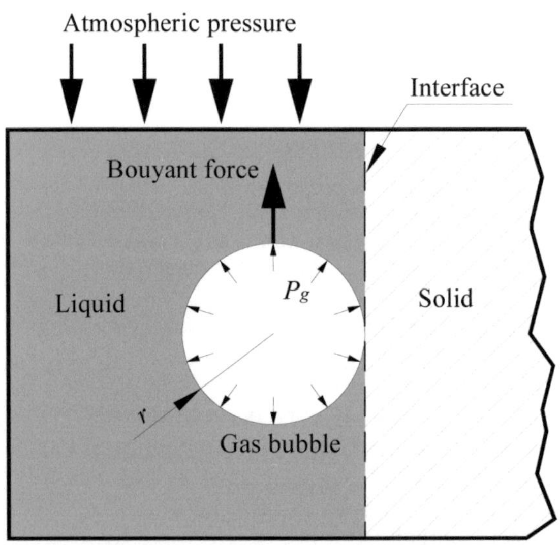

Figure 9. Gas bubble growth due to the diffusion in liquid phase.

http://dx.doi.org/10.5772/58418

The critical radius r_c for the separation of the bubble from the solid-liquid interface can be determined by means of the following expression [11]:

$$r_c = 7.5 \times 10^{-4} \kappa \left[\frac{2\sigma^*}{g(\rho_l - \rho_g)} \right]^{1/2} \quad (5)$$

where κ is the detaching angle of the bubble, σ* the interfacial energy between liquid and gas (~ 1 Jm^{-2} for the majority of the gas-metal systems), g the constant gravity force and ϱ_l-ϱ_g the difference between liquid and gas densities. Considering that densities of liquid aluminum and hydrogen at melting point of aluminum are approximately 2380 kgm^{-3} and 0.0256 kgm^{-3}, respectively, and κ=100°, the r_c is roughly 700 μm. It is to say that gas bubbles formed during welding process have to be greater than 700 μm to overcome the surface tension of the solid-liquid interface. Additionally, a low solidification rate is needed to allow the gas bubble coalescence and reach this critical value. However, in a real situation the cooling thermal cycle of the fusion zone in welding, is very fast, as demonstrated in [8] (Figure 10), and porosity formation is present as shown in Figure 11.

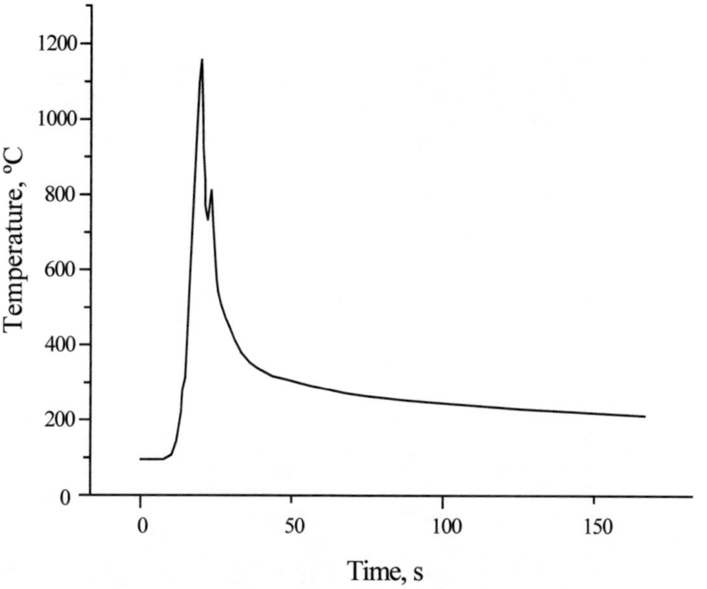

Figure 10. Profile of the temperature measured in the weld pool on a 6061-T6 aluminum alloy GMAW weld [8].

Liquation in welds of aluminum alloys can occur in the partially melted zone (PMZ). The PMZ is the region outside the fusion zone where grain boundary liquation occurs during welding. Figure 12 shows a portion of the PMZ in GMAW of 6061-T6 aluminum made with high silicon content filler metal (ER4043).

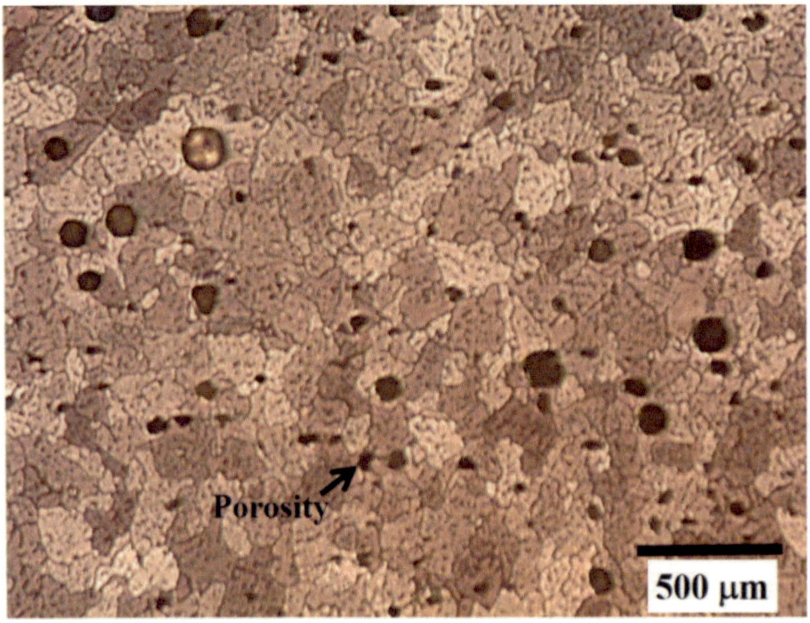

Figure 11. Weld pool porosity on a 6061-T6 aluminum alloy welded by GMAW.

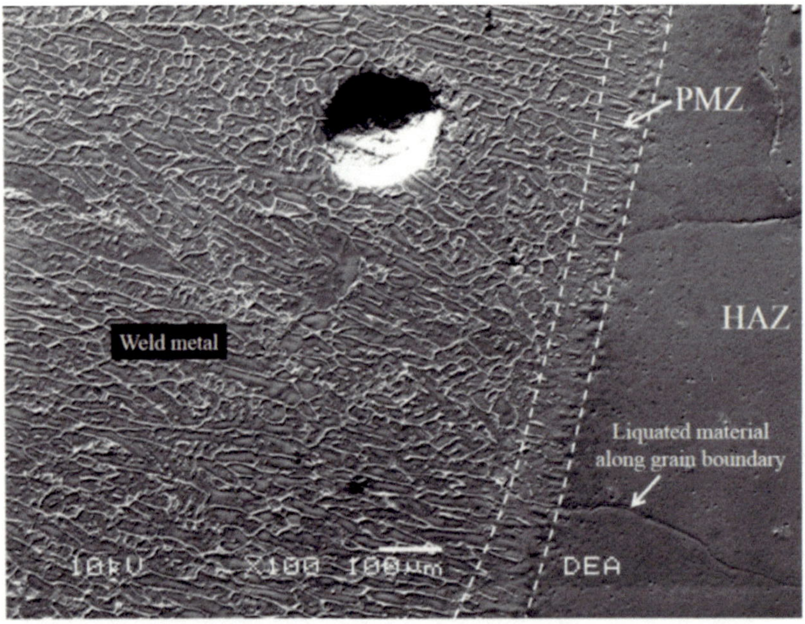

Figure 12. Microstructure of a 6061-T6 aluminum, welded by GMAW with ER4043 filler metal, showing the PMZ, and grain boundary liquation.

http://dx.doi.org/10.5772/58418

The liquation phenomenon occurs along grain boundary, although it can be presented in the grain interior. When liquation is presented, cracking can occur along grain boundary because of the tensile strains generated during welding. The weld metal composition is determined by base metal and filler metal compositions, as well as, the dilution ratio. The dilution ration is related with the amount of filler metal diluted with the base metal to form the weld metal. Metzger [12] observed liquation cracking in full penetration, gas tungsten arc welding (GTAW) on a 6061 alloy made with Al-Mg filler metal at high dilution ratios, but not in similar welds made with Al-Si filler metals at any dilution ratio. Huang et al. [13] have conducted studies on liquation cracking in the PMZ of full penetration welds of a 6061 alloy. They found that liquation cracking occurred in GMAW welds when an ER5356 (Al-Mg) filler metal is employed but not with an ER4043 (Al-Si).

2. Mechanical behavior of aluminum alloys welds

2.1. Indentation

The hardness of a material represents the plastic deformation resistance by indentation. The hardness number *H*, is given by the ratio between the applied load *P*, and a representative area *A*, of the residual indent:

$$H = \frac{P}{A} \tag{6}$$

For usual indentation, the Vickers hardness number *HV*, using a pyramidal square based indenter is calculated considering the true contact area. The relation of *HV* is given by:

$$HV = 1.8544 \frac{P}{d^2} \tag{7}$$

where, sometimes, *HV* is expressed in MPa if *P* is given in N and *d*, the indent diagonal, in mm. But usually *HV* is given as a number and the conditions of load used.

Concerning instrumented indentation tests (IIT), which allow to plot a load-depth curve, the calculation of a hardness number can use the maximum distance (maximum indentation depth h_m, reached by the indenter during the indentation test) or the contact depth which is the indentation depth h_c, taking into account the deformation of the indent under load and calculated using the method of Olive and Pharr [14].

Classical indentation has been used to determine the hardness evolution in precipitation hardening aluminum alloys welds [8, 15, 16]. Ambriz et al. [8] determined Vickers microhardness profiles and mapping representation in welds of 6061-T6 aluminum alloy (Figure 13).

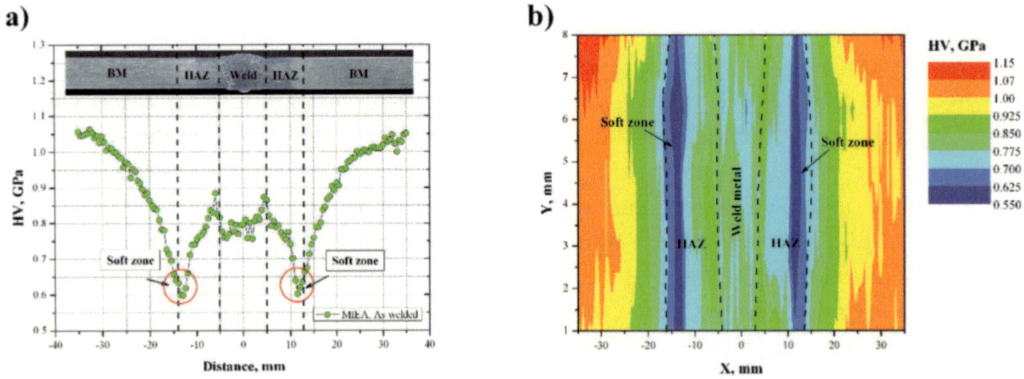

Figure 13. a) Vickers hardness profile determined with 1 N of applied load throughout the welded joint and b) Vickers hardness map over the welded joint [8].

A significant difference for the hardness of weld metal, and HAZ with respect to base material was observed. This indicates that mechanical properties after welding will be different. It should be noted a soft zone formation in both sides of the welded joint, the hardness decrease in the soft zone is around 43% with respect to base material. This characteristic is the result of the thermodynamic instability of the β'' needle-shaped precipitates (hard and fine precipitates) promoted by the high temperatures reached during the welding process. Indeed the temperatures reached during the welding process are favorable to transform the β' phase, rod-shaped according to the transformation diagram for the 6061 alloy (Figure 6).

Considering the hardness heterogeneity of the welded joints, instrumented indentation tests (IIT) was performed in base metal, weld metal and HAZ (soft zone). Figure 14, shows the evolution of the applied load as a function of indentation depth for 6061-T6 and 7075-T651 aluminum alloys welds.

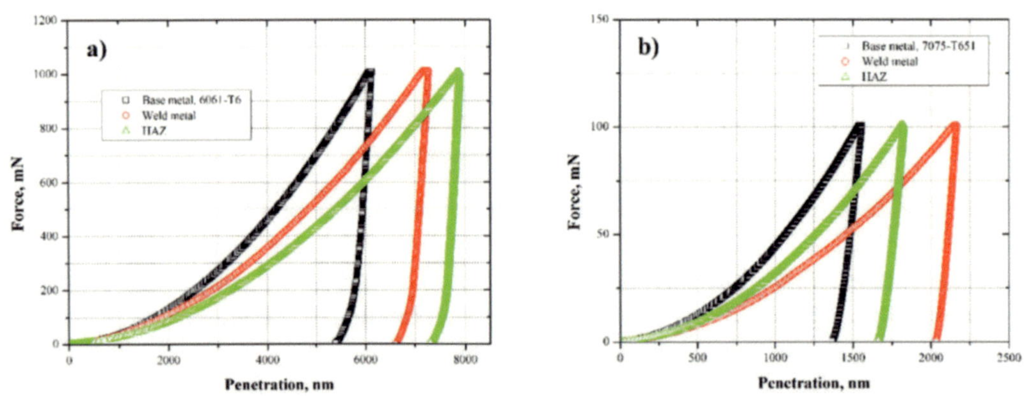

Figure 14. Load-depth curves for a) 6061-T6 aluminum alloy (1 N of applied load) and b) 7075-T651 aluminum alloy (0.1 N of applied load).

Moreover, from the instrumented indentation tests it is possible to calculate the elastic modulus which is deduced by the inverse of the unloading curve (1/S) as a function of the inverse of the contact indentation depth (1/h_c) (equation 6).

$$\left(\frac{dh}{dP}\right) = C_f + \sqrt{\frac{\pi}{24.5}} \cdot \frac{1}{2(\beta \cdot \gamma) \cdot E_R} \cdot \frac{1}{h_c} \tag{8}$$

where $\left(\frac{dh}{dP}\right)$ is the inverse of the contact stiffness, C_f the frame compliance of the indentation instrument, β geometrical factor introduce in the model of Oliver and Pharr [14] to take into account the indenter shape and γ the factor introduced by Hay et al. [17] in the model of Oliver and Pharr for taking into account the approximation in the Hertz's contact analysis and E_R the reduced modulus.

To calculate E_R it is needed to determine a corrective factor due to Hay et al. [17]. This coefficient is only dependent on the Poisson's ratio, then considering a constant value of ν_m=0.3 for any region of the welded joint (base material, weld metal and HAZ), it is obtained a value of 1.067 for γ. In this condition, the slope is only linked to the reduced modulus by 0.1653/E_R. Considering 1140 GPa and 0.07 for the elastic modulus and the Poisson's ratio for the indenter material, respectively, it is possible to determine the elastic modulus of the different zones of the welded joints (Table II), by means of the following equation:

$$E_R = \left(\frac{1-\nu_m^2}{E_m} + \frac{1-\nu_i^2}{E_i}\right)^{-1} \tag{9}$$

where E_m and ν_m are the elastic modulus and Poisson's ratio of the material and E_i and ν_i the elastic modulus and Poisson's ratio of the indenter, respectively.

Additionally, the yield strength σ_y and hardening exponent n can be obtained by indentation tests (Table II) as suggested by Ambriz et al. [18], by means of the following expression:

$$P = P_0 + 26.43\sigma_y \left\{ \frac{2}{3} + \frac{1}{2}\left(\frac{1}{3}\frac{E}{\sigma_y}\cot(\lambda)\right)^n + \frac{2}{3n}\left[\left(\frac{1}{3}\frac{E}{\sigma_y}\cot(\lambda)^n\right) - 1\right] \right\} h^2 \tag{10}$$

where P is the applied indentation load, P_0 shift indentation load, E elastic modulus, h true indentation depth into material.

Instrumented indentation test allows to determine the properties given in Table II. Yield strength values determined by instrumented indentation are similar to those obtained by micro-traction test (Figure 16). It means that it is possible to characterize a local zone in a welded material by instrumented indentation where it is not possible by global test (tensile test).

Material	E_{IIT} (GPa)	σ_y (MPa)	n	P_0 (N)
Base metal	98.6	266	0.46	0.015
Weld metal	78.0	133	0.48	0.022
HAZ	87.4	106	0.46	0.016

Table 2. Mechanical properties obtained by instrumented indentation test for a 6061-T6 aluminum alloy welds [18].

On the other hand, the elastic modulus obtained by instrumented indentation does not correspond with those reported in tension or compression tests for aluminum alloys (68-72 GPa) and it is not possible to establish a clear difference between weld metal and HAZ. In fact in a recent study, Chicot et al. [19] established that the elastic modulus obtained by instrumented indentation corresponds to a bulk modulus. This was explained by the fact that for the case of indentation, the elastic and plastic strain is triaxial, whereas in tensile test, the strain used to determine the elastic modulus is uniaxial.

2.2. Tensile properties

As possible to deduce from indentation test (Figure 13), the tensile mechanical properties of aluminum alloys welds hardened by precipitation, are not homogeneous along the welded joint. Tensile properties in welds obtained by several welding processes of this alloys have been studied. For instance, V. Malin [15] studied the relation between weld thermal cycles and the microstructural transformation with tensile properties of a 6061-T6 alloy welded by GMAW. In their research, tensile tests samples were taken from the welded joint as shown in Figure 15. It is to say, a global structure effect was considered during tensile test.

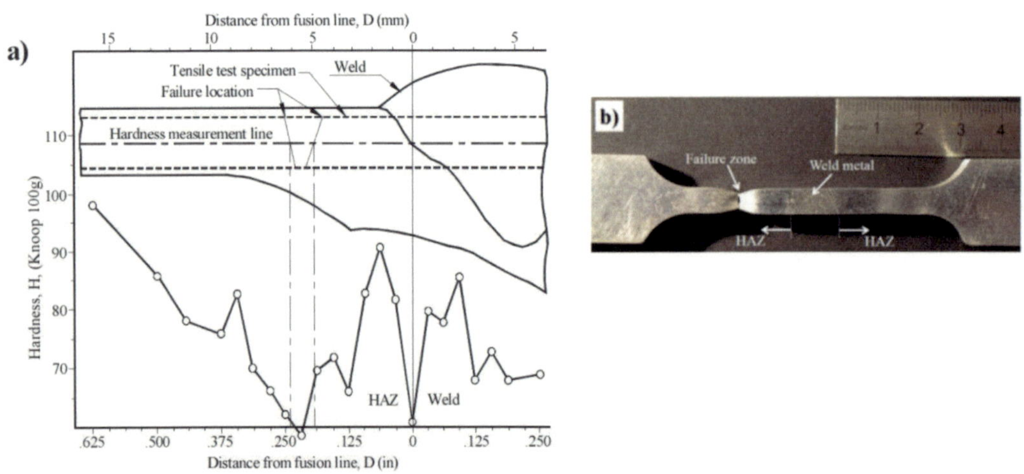

Figure 15. (a) Effect of hardness on HAZ failure location in 6061-T6 tensile specimen [15], and b) tensile sample showing the failure zone on a 6061-T6 aluminum alloy.

http://dx.doi.org/10.5772/58418

The failure zone after tensile test was localized in the HAZ (soft zone) where the hardness of the welded joint is minimal, this result is in agreement with Ambriz et al. [20].

Considering the hardness profile evolution, the true stress-strain curves in weld metal and HAZ (soft zone) of a 6061-T6 alloy welds were determined by means of micro-traction test [18]. The individual behavior is presented in Figure 16, as well as, its respective comparison with base metal.

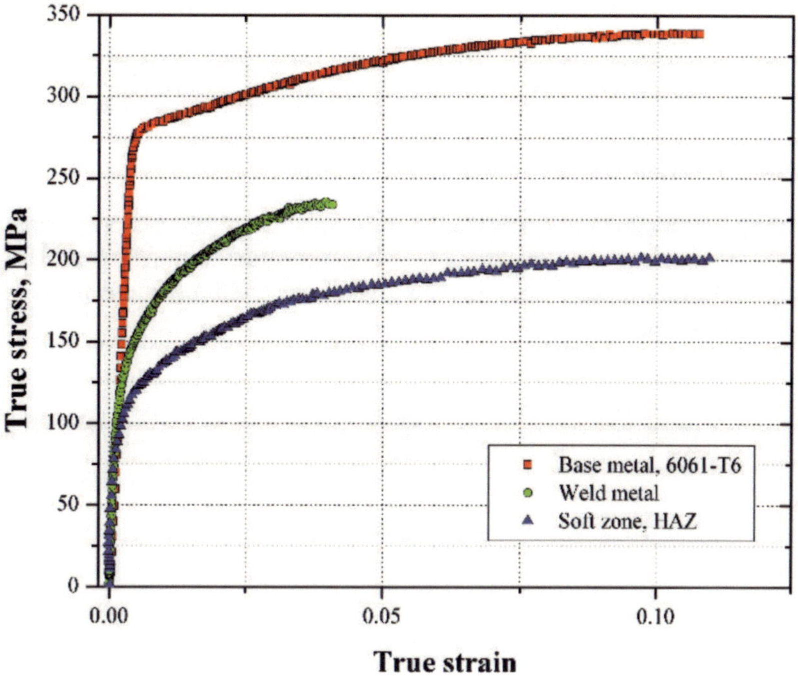

Figure 16. True stress-strain curves for 6061-T6 alloy, weld metal and HAZ.

The HAZ presents a reduction of the tensile strength with respect to base metal and weld metal of around 41 and 19%, respectively. This aspect was related to the over-aging phenomenon and it is explained in terms of the microstructural transformation (Figure 5), and precipitation sequence. Although, weld metal shows higher tensile strength than HAZ, a lower ductility is observed for weld metal. This characteristic was attribute to the porosity formation during the solidification and the high silicon content of the filler metal (ER4043) which, when mixed with the melted base metal, leads to a microstructure of eutectic silicon, which is a brittle phase that adversely affects the tensile mechanical properties of the welded joint. A summary of tensile mechanical properties for 6061-T6 and 7075-T651 aluminum alloys welds are presented in Table III.

Material	E (MPa)[a]	σ_y (MPa)	σ_u (MPa)[b]	ε (%)[c]	H (MPa)[d]	n^d
6061-T6	68	279	310	14.0	408	0.08
6061-T6 (weld metal, ER4043)	68	151	226	4.00	464	0.20
6061-T6 (HAZ, soft zone)	68	120	183	13.0	300	0.16
7075-T651	72	530	568	8.00	794	0.08
7075-T651 (weld joint, ER5356)	68	165	260	2.80	677	0.25

[a] Best linear fitting from mean stress-strain curves; [b]Obtained from $\sigma = \frac{P}{A_0}$; [c] obtained from $\varepsilon = \frac{l_f - l_0}{l}$, [d] obtained by Ramber-Osgood equation.

Table 3. Tensile mechanical properties of 6061-T6 and 7075-T651 aluminum alloys welds (GMAW).

2.3. Fatigue of aluminum alloys welds

Fatigue or fatigue damage is the consecutive modification of the materials properties with respect to the application of a cyclic stress, which can conduct to the fracture. Under uniaxial cyclic loading conditions it is possible to distinguish a stress ratio R determined by a maximal and minimal stress as follow:

$$R = \frac{\sigma_{\min}}{\sigma_{\max}} \tag{11}$$

As a function of σ_{max} and σ_{min}, we can obtain the constant component of the stress (amplitude stress σ_a) and the mean value of the stress (mean stress σ_m) by different loading conditions (Figure 17).

Some practical applications involve cyclic loading at a constant amplitude, but irregular loads as a function of time are commonly encountered. In this case, we will discuss some results in terms of a constant amplitude loading. The simplest fatigue test consists of subjecting a specimen to a cycling loading (different levels of stress amplitude σ_a), at a constant frequency and measured the number of cycles to failure N_f. The representation of stress level as a function of N_f gives the *S-N* curve or Wöhler curve. A schematic representation of a Wöhler curve is shown in Figure 18.

In Figure 18, it is possible to identify different domains: (i) Low cycle fatigue. In this case a high stress level is applied on the sample (normally over the yield strength of the material). Because of the high deformation during the test, the number of cycles to failure tends to be lower (10^2 to 10^4). (ii) High cycle fatigue. This is related with an elastic behavior on a macro scale of the sample, i.e. the stress level is not higher than the yield strength of the material. The failure is expected for a large number of cycles, for instance, more than 10^5. In fact the boundary between low and high cycle fatigue is not well defined by a specific number of cycles. The most important difference is that low cyclic fatigue is associated with macro-plastic deformation on

http://dx.doi.org/10.5772/58418

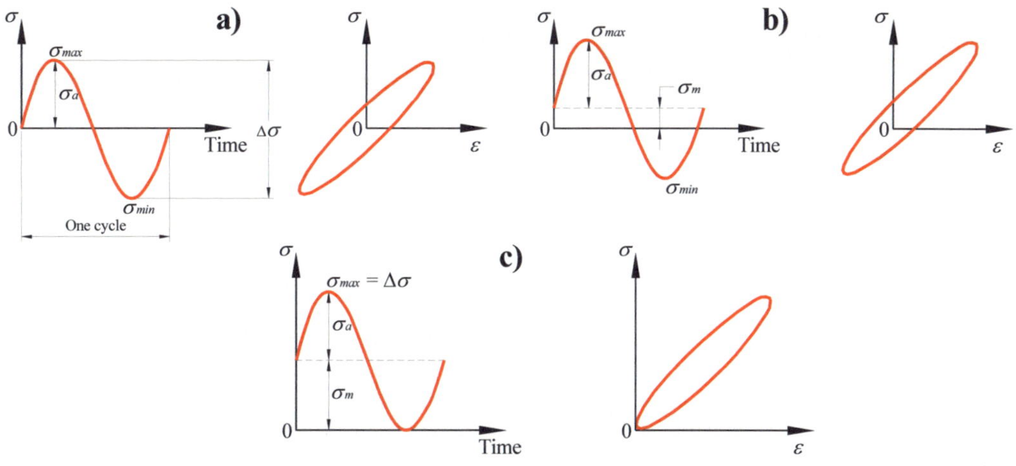

Figure 17. Solicitation loading in uniaxial fatigue, a) completely reversed stressing (σ_m=0), b) asymmetric repeated stressing ($\sigma_m \neq 0$), and c) zero to tension stressing (σ_{min}=0).

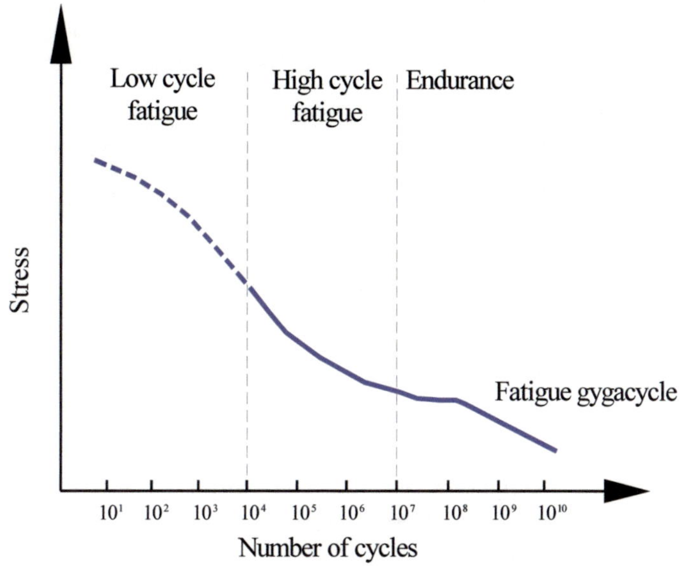

Figure 18. Representation of a Wöhler curve, and different fatigue domains [21].

each loading cycle. When the stress level in high cycle fatigue is applied, a fatigue limit or endurance limited is presented, which is represented by an asymptote in the Wöhler curve. In some metallic materials it could be obtained when the number of cycles is in the order of 10^6 to 10^7. (iii) Fatigue gigacycle. This domain corresponding with a very high number of cycles and it has been observed that fatigue limit tends to decrease when the number of cycles increases.

It is well known that fatigue damage is a surface phenomenon as indicated by Forsyth [22], who determined the presence of reliefs linked to the formation of localization deformation bands named persistent bands. The surface topography is traduced by the formation of intrusions and extrusions as shown schematically in Figure 19.

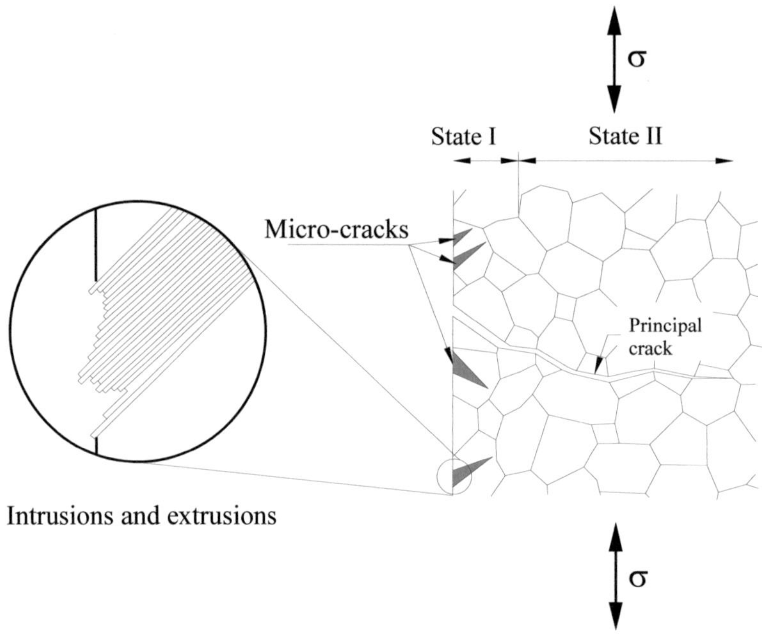

Figure 19. Intrusions and extrusions formation in the free surface due to the alternating slip, micro-cracks nucleation, and principal crack formation from micro-cracks [21].

For a uniaxial tensile test, these bands resulting in the formation of micro-cracks (state I in Figure 19), which are orientated at 45 degrees with respect the traction axe. Only certain grains are affected by the formation of those bands. The persistent bands orientation and the formation of cracks on the state I, are important in the case of uniaxial and multiaxial loading. Brown and Miller [23, 24] introduced a useful notation in multiaxal loading for facets A and B, which are schematized in Figure 20.

The type B facets provides a shearing vector which enters into the material, and they are more dangerous than type A facets, from which the shearing vector is tangent to the free surface of the sample. The intrusions and extrusions formation associated to the slip persistent bands, as well as, the micro-propagation of cracks in the state I are of interest at a distance of the grain size (small fatigue cracks). Thus, considering that micro-cracks are related with the crystallography aspect, once the crack encounters the first grain boundary it begins to bifurcate according to state II and the propagation at a perpendicular direction of the principal stress is obtained.

http://dx.doi.org/10.5772/58418

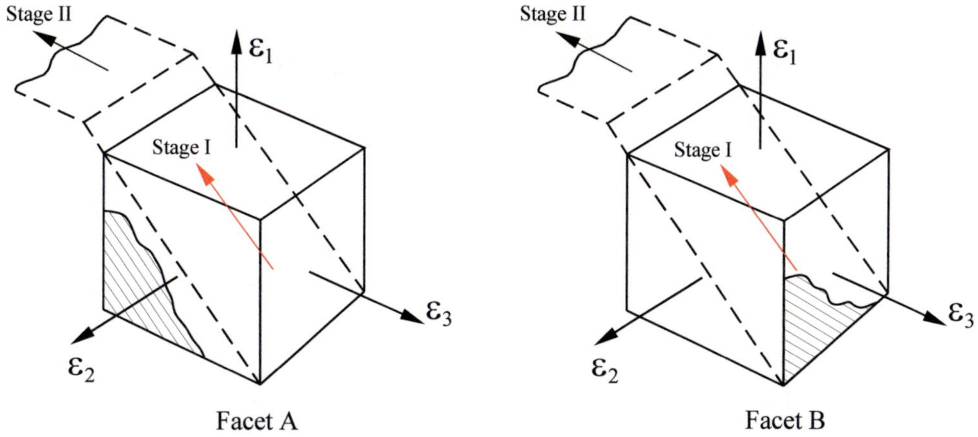

Figure 20. Directional aspect of the fatigue damage. Importance of the stress field orientation with respect to the plans surface, and the free surface of the material (hatch zones) [23, 24].

Additionally to the fatigue damage mechanism mentioned previously, in the case of welding, the stress concentration factor due to the geometry of the welding bead has a special importance. In this sense, Ambriz et al. [20] has been quantified the effect of the welding profile generated by modified indirect electric arc (MIEA) technique on the fatigue life of a 6061-T6 aluminum alloy. In order to determine the stress concentration factor K_t in MIEA welds, a characteristic welding profile was measured as specified in Figure 21.

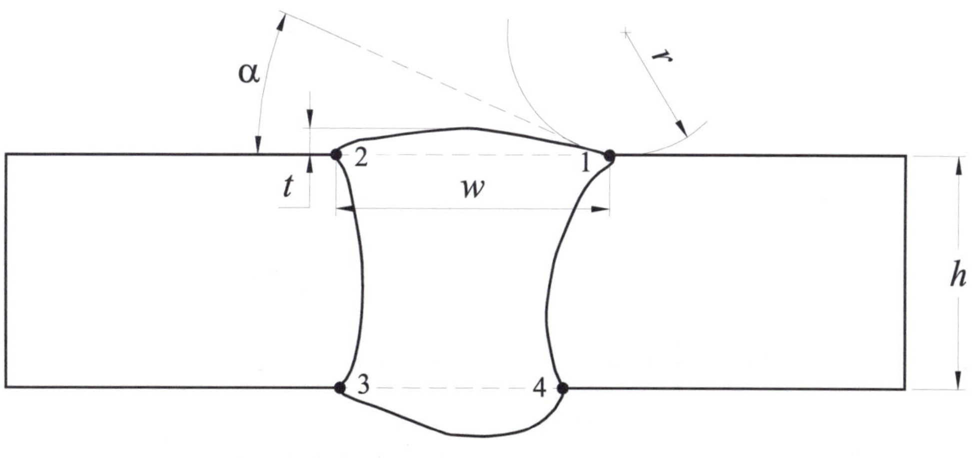

Figure 21. Stress concentration points in MIEA welds and their corresponding dimensions. α=angle formed by the weld reinforcement with plates, *r*=notched radius, *t*=height of the weld reinforcement, *w*=width of the welding profile and *h*=thickness of the plates.

Subsequently, uniaxial fatigue test a cyclic loading with a sinusoidal wave form at a frequency of 35 Hz and load ration R=0.1 was applied in atmospheric air at room temperature. The fatigue limit (77 MPa) was calculated employed the Locati method, and Wöhler curve (Figure 22) was plotted between 77 and 110 MPa.

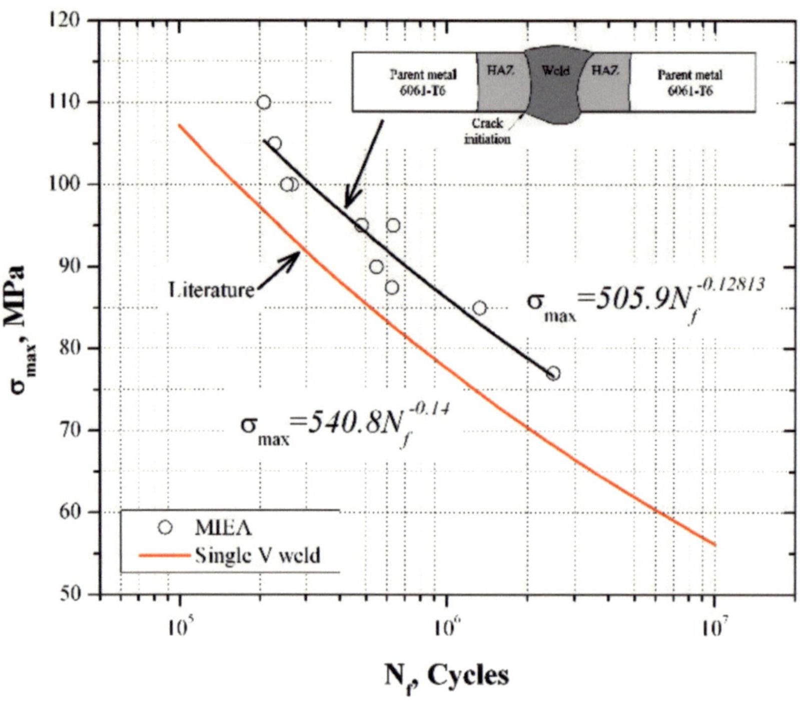

Figure 22. Wöhler curve for 6061-T6 aluminum alloy welds by MIEA and data found in literature [25] for the single V groove joint.

Figure 22 shows the maximum stress σ_{max}, as function of number of cycles to failure N_f, obtained from the fatigue test of the welded joints. The experimental results were fitted according to the following expression:

$$\sigma_{\max} = AN_f^b \tag{12}$$

where A, and b are experimental values determined by fitting curve.

Regarding the geometry of the welding profile, comparison of the fatigue performance exhibited by MIEA welds with the results reported in the literature [25] for a single V joint configuration for the same aluminum alloy shows a significant improvement in fatigue life for the MIEA welded samples. The K_t effect on the crack initiation and failure is shown in Figure 23.

Figure 23. Typical macrograph of the fatigue failure on MIEA welds.

Considering that, if the loading condition promotes the formation of a principal crack (Figure 19), it will grow according to the power law region as shown schematically in Figure 24.

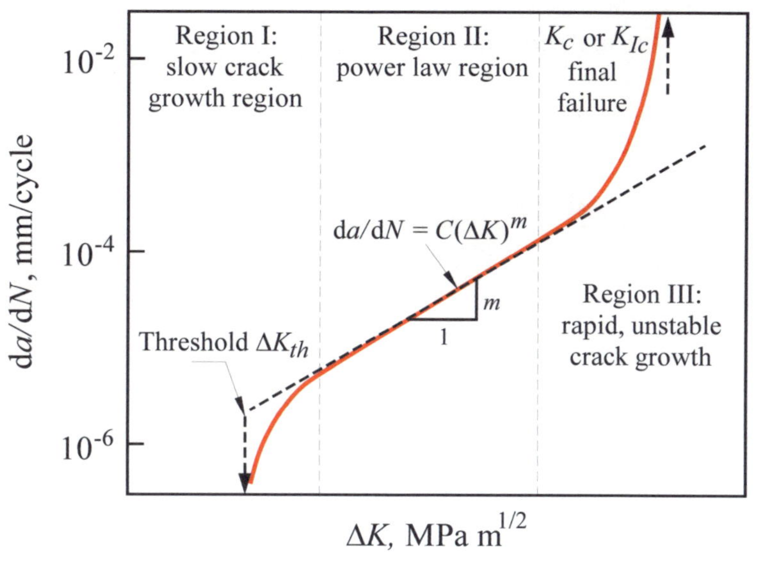

Figure 24. Fatigue crack growth regimes as function of Δ*K*.

The fatigue crack growth rate d*a*/d*N*, as function of stress intensity factor range Δ*K* in different zones of the welded joint (base metal, weld metal and HAZ) has been studied previously [26, 27]. The results were conducted on compact type specimens (CT) applying a constant amplitude cycling load with a sinusoidal wave form at a frequency of 20 Hz, a load ratio *R* of 0.1 and a load range of 2.5 kN in atmospheric air at room temperature. The Δ*K* was computed by means of the following equation:

$$\Delta K = \frac{\Delta P\left(2+\frac{a}{W}\right)}{BW^{1/2}\left(1-\frac{a}{W}\right)^{3/2}}\left[0.886+4.64\frac{a}{W}-13.32\left(\frac{a}{W}\right)^2+14.72\left(\frac{a}{W}\right)^3-5.6\left(\frac{a}{W}\right)^4\right] \quad (13)$$

where a is the crack length of the sample (initial crack of 8 mm), W the width for the crack propagation, and B the thickness of the sample (5 mm in this case).

Considering the stable crack growth propagation region shown in Figure 24, the experimental results of a were plotted in da/dN versus ΔK graphs according to the following expression:

$$\frac{da}{dN} = C(\Delta K)^m \quad (14)$$

where C and m are constants obtained from the fitting curve. Table IV summarizes the best fitted values for C and m, and their correlation factor.

Material	***C***	***m***	R^2
6061-T6 (rolling)	5.00×10^{-7}	2.426	0.89
6061-T6 (transverse)	3.97×10^{-8}	3.320	0.97
Weld metal	2.63×10^{-14}	8.550	0.92
HAZ	3.77×10^{-9}	6.650	0.89

Table 4. Fitting constants obtained from experimental values (Figure 25) [27].

Figure 25, shows the fatigue crack growth for base metal (6061-T6) in the rolling and transverse to rolling direction.

This graph shows that the microstructural characteristics (anisotropy) does not have an important influence in terms of fatigue crack growth as could be expected, taking into account that yield strength in rolling direction is higher than transverse direction. However, this is not the case for weld metal and HAZ (Figures 25b and c), in which the crack tends to propagate faster than that in base metal. In the case of weld metal (Figure 25 b), the faster crack growth rate in comparison with base metal is related to the low toughness due to the high silicon (~ 5.5 wt. percent) content provided by the filler metal during welding. Similarly, for the HAZ it is possible to observe that the crack growth is faster than base metal, aspect which is attribute to the microstructural transformation of fine needle shape precipitates β'' into coarse bar shape precipitates β' produced by the thermal effect.

http://dx.doi.org/10.5772/58418

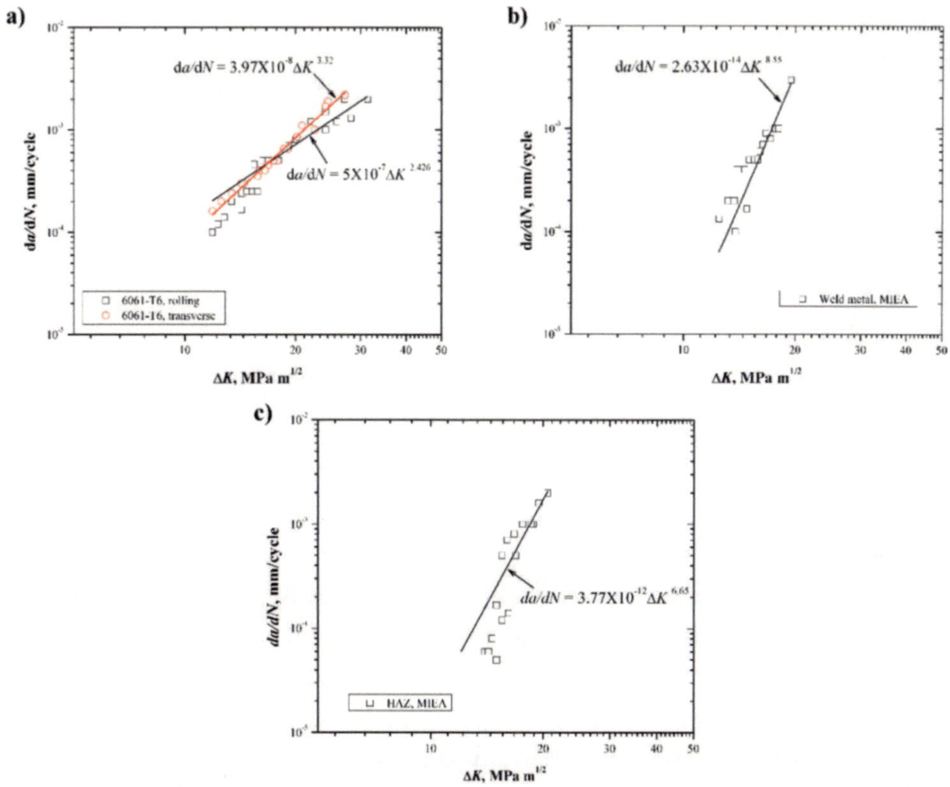

Figure 25. Fatigue crack growth rate as function of Δ*K*, a) 6061-T6 base metal, b) weld metal, and c) HAZ.

3. Conclusion

Mechanical behavior in welds of precipitation hardened aluminum alloys are still under development and the softening phenomena in the heat affected zone should be better understood. Valuable information could be obtained by the precise understanding of the weld thermal cycles in conjunction with the C transformation curve and its microstructural effect in mechanical properties. In this sense, our research group is conducting experiments to control the weld thermal cycle by means of localized chillers and heaters in the fusion zone and heat affected zone to observe the mechanical properties evolution of the welded joints.

Author details

R.R. Ambriz and D. Jaramillo

Instituto Politécnico Nacional CIITEC-IPN, Cerrada de Cecati S/N Col. Sta. Catarina, Azcapotzalco, DF, México

References

[1] Gladman T. Precipitation Hardening in Metals. Materials Science and Technology 1999;15 30-36.

[2] Orowan E. Internal Stress in Metals and Alloys; 1948.

[3] Ashby MF. Oxide Dispersion Strengthening. Gordon and Breach; 1958.

[4] Gladman T. The Physical Metallurgy of Microalloyed Steels. The Institute of Materials; 1997.

[5] ASM. Properties and Selection: Non Ferrous Alloys and Special Purpose Materials. ASM International; 1992.

[6] Askeland D., Fulay P and Wright W. The Science and Engineering of Materials; 2010.

[7] Edwards G., Stiller K., Dunlop GL. and Couper MJ. The precipitation sequence in Al-Mg-Si alloys. Acta Materialia 1998;46 (11) 3893-3904.

[8] Ambriz RR., Barrera G., García R. and López VH. Effect of the Weld Thermal Cycles of the Modified Indirect Electric Arc on the Mechanical Properties of the AA6061-T6 Alloy. Welding International 2010;24 (4) 42-51.

[9] Myhr OR., Grong O., Fjaer HG. and Marioara CD. Modelling of the Microstructure and Strenght Evolution in Al-Mg-Si Alloys During Multistage Thermal Processing. Acta Materialia 2004;52 4997-5008.

[10] Matters G. The Welding of Aluminum and its Alloys. CRC; 2002.

[11] Grong O. Metallurgical Modelling of Welding. The Institute of Materials; 1997.

[12] Metzger GE. Some Mechanical Properties of Welds in 6061 Aluminum Alloy Sheet. Welding Journal 1967;46 (10) 457-469.

[13] Huang C. and Kou S. Liquation Cracking in Full Penetration Al-Mg-Si Welds. Welding Journal 2004;4 111-122.

[14] Oliver WC. and Pharr GM. An Improved Technique for Determining Hardness and Elastic Modulus Using Load and Displacement Sensing Indentation Experiments. Journal of Materials Research 1992;7 (6) 1564-1583.

[15] Malin V. Study of Metallurgical Phenomena in the HAZ of 6061-T6 Aluminum Welded Joints. Welding Journal 1995;9 305-318.

[16] Liu G., Murr LE., Niou CS., McClure JC. and Vega FR. Microstructural Aspects of the Friction Stir Welding of 6061-T6 Aluminum. Scripta Materialia 1997;37 (3) 355-361.

[17] Hay JC., Bolshakov A. and Pharr GM. Critical Examination of the Fundamental Relations Used in the Analysis of Nano-Indentation Data. Journal of Materials Research 1999;14 (6) 2296-2305.

[18] Ambriz RR., Chicot D., Benseddiq N., Mesmacque G. and de la Torre SD. Local Mechanical Properties of the 6061-T6 Aluminum Weld Using Micro-Traction and Instrumented Indentation. European Journal of Mechanics A/Solids 2011;30 307-315.

[19] Chicot D., Roudet F., Zaoui A., Louis G. and Lepingle V. Influence of Visco-Elastic-Plastic Properties of Magnetite on the Elastic Modulus: Multicyclic Indentation and Theoretical Studies. Materials Chemistry and Physics 2010;119 75-81.

[20] Ambriz RR., Mesmacque G., Ruiz A., Amrouche A. and López VH. Effect of the Welding Profile Generated by the Modified Indirect Electric Arc Technique on the Fatigue Behavior of 6061-T6 Aluminum Alloy. Materials Science and Engineering A 2010;527 2057-2064.

[21] Pineau A. and Bathias C. Fatigue des Matériaux et des Structures 1. Lavoisier; 2008.

[22] Forsyth PJE. Some Metallographic Observations on the Fatigue of Metals. Journal of the Institute of Metals 1951;80 181.

[23] Brown MW. and Miller KJ. A Theory for Fatigue Failure Under Multiaxial Stress-Strain Conditions. Proceedings of the Institution of Mechanical Engineerings 1973;187 745-755.

[24] Miller KJ. Metal Fatigue-Past, Current and Future. Proceedings of the Institution of Mechanical Engineerings 1991;205 1-14.

[25] ASM. Fatigue and Fracture. ASM Iternational; 1996.

[26] Moreira PMGP., de Jesus AMP., Ribeiro AS. and de Castro PMST. Fatigue Crack Growth in Friction Stir Welds of 6082-T6 and 6061-T6 Aluminium Alloys: A Comparison. Theoretical and Applied Fracture Mechanics 2008;50 81-91.

[27] Ambriz RR., Mesmacque G., Ruiz A., Amrouche A., López VH. and Benseddiq N. Fatigue Crack Growth Under a Constant Amplitud Loading of Al-6061-T6 Welds Obtained by Modified Indirect Electric Arc Technique. Science and Technology of Welding and Joining 2010;15 (6) 514-521.

Chapter 3

Mechanical Behaviour of Commercial Aluminium Wrought Alloys at Low Temperatures

R. Schneider, B. Heine and R.J. Grant

Additional information is available at the end of the chapter

http://dx.doi.org/10.5772/58362

1. Introduction

For a long time, aluminium wrought alloys have been used as the predominant material for the sheet metal processing industry. Aluminium alloys not only show a high affinity for passivation which results in corrosion resistance in many environments [1], but also indicate high specific strengths (strength versus density) and rigidity (Young´s modulus versus density). The most commonly used aluminium alloys in aircraft applications of the non-precipitation hardenable groups are EN AW-3xxx, EN AW-5xxx and of the precipitation hardenable groups EN AW-2xxx, EN AW-6xxx and EN AW-7xxx. The advantages of the EN AW-2xxx group are good mechanical properties which remain at higher temperatures, whereas the EN AW-7xxx groups show good mechanical properties at lower temperatures and high stress corrosion resistivity. A negative aspect is, however, that aluminium alloys have a very low formability when compared with modern steels. Figure 1 illustrates the fracture elongation versus yield strength of aluminium wrought alloys and deep drawing steels in optimised hardened condition. Aluminium wrought alloys with the designation EN AW-7xxx series in peak-aged condition show yield strength values comparable with the micro-alloyed steels; although the fracture elongation differs slightly in favour of the steels.

Yield strength values from the wrought alloys of EN AW-6xxx group reach only about 50% of the strength which can be obtained from the EN AW-7xxx group at ambient temperatures; both aluminium wrought alloys are precipitation hardenable in peak-aged condition. The fracture elongation of the 6xxx and 7xxx group can be identified to be approximately half of the value which can be obtained using bake-hardening steels. Yield strength values from non-heat-treatable, and therefore only strain-hardenable, aluminium alloys of the EN AW-5xxx group are slightly higher when present in highly strain-hardened condition (e.g. H18). But their fracture elongation values in an annealed condition are equivalent to those exhibited by

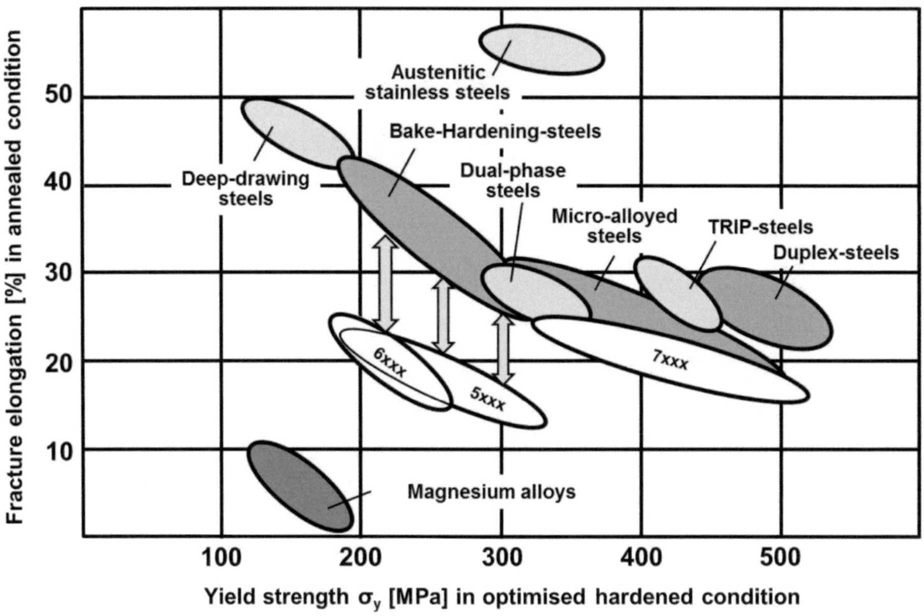

Figure 1. Fracture elongation for aluminium wrought alloys of the EN AW-5xxx, EN AW-6xxx and EN AW-7xxx series at T=298 K and deep-drawing steels at T=298 K. Yield strength in an optimised hardened condition for all mentioned groups at T=298 K.

the EN AW-6xxx group. So, ways of enhancing the fracture elongation of wrought aluminium alloys to levels that are at least comparable with those of deep-drawing steels and additionally possess corresponding strength values are sought. Forming aluminium alloys below room temperature would be an approach, where strain hardening losses and precipitation processes are not an issue. In this regard there is no danger of a reduction in strength through the mechanism of recovery, recrystallization and aging which is in most cases induced by forming at higher temperatures. Furthermore, in previous investigations during tensile testing it was observed that in certain aluminium alloys, which were present in various heat treatment conditions, the Lüders and the Portevin-LeChatelier (PLC) effect occurs at room temperature [2] [3]. This PLC effect not only causes a bad surface quality on sheet metals, but also creates a reduction of the uniform elongation and fracture elongation [2]. It is expected that in tensile testing at low temperatures the negative impact caused by the PLC effect can be reduced and therefore the values of uniform elongation and fracture elongation can be increased.

2. Portevin-LeChatelier (PLC) Effect

2.1. Phenomena of the PLC effect

Considering a tensile testing specimen, consisting of an isotropic ductile material, it can be observed that straining in the direction of the principal normal stress is accompanied by a cross-sectional reduction in width and thickness. In the region of uniform elongation the prismatic geometry of the specimen remains, since a *random localisation* stabilises itself according to the Considère-criteria. Leaving this region of the stress-strain curve and moving into the region which signifies the region of necking behaviour a localised and random reduction in width and thickness results in an overall reduction in cross-section. It is of note that for a tensile specimen, where the width is much greater than the thickness, a reduction in the cross-section only appears at a volume of material inclined at an angle of approximately 55° measured in direction of the principal normal stress. Localised necking predominantly in thickness will finally fracture the specimen when constricted [4] [5]. Testing an annealed aluminium alloy with the designation of EN AW-5xxx shows that a cross-sectional reduction occurs in fact before reaching the necking area. Figure 2 a) shows localised necking over the whole specimen length.

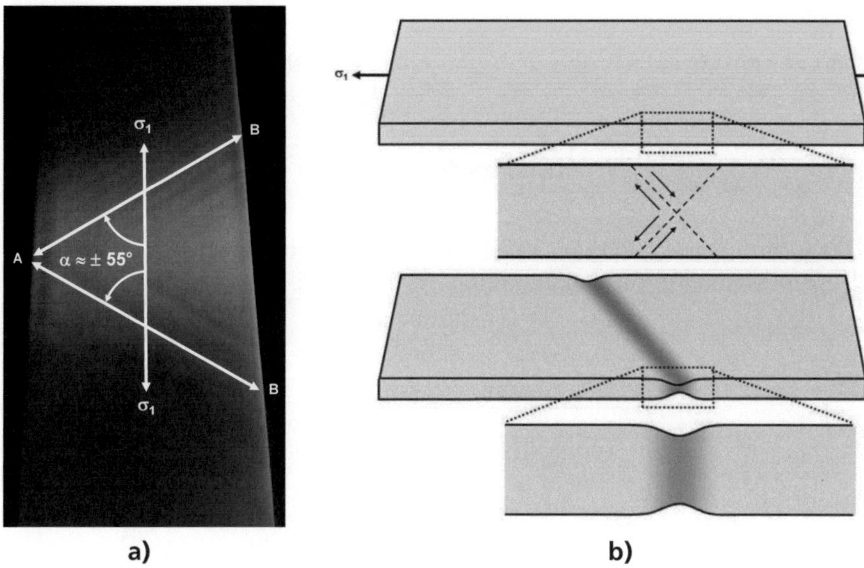

Figure 2. a) Cross-sectional reduction of EN AW-5xxx specimen in annealed condition which shows localised necking at an angle of around 55° measured in direction of principal normal stress before reaching the necking area. b) Schematic cross-sectional reduction in thickness co-occurring with shear deformations in several planes.

In this deformation state it can be observed that only localised necking occurs ($\overline{AB}$) and there is no reduction in width over the whole specimen length ($\overline{BB}$) which would be denoted as diffuse necking [6]. This is **in contrast** to most metallic materials, where localised through-thickness necking arises e.g. on aluminium alloy EN AW-5xxx in an annealed condition, but not over the whole length. As a consequence, a pattern of localised through-thickness striations can be observed oriented at around ± 55° to the direction of the principal normal stress (Figure 2 a)) [2] [3]: these are also known as PLC bands. Figure 2 b) illustrates a formation of this localised affect which is co-occurring with shear deformations in several planes. The condition of the volume of material where localised necking arises is that no strain be imposed on the non-deforming material adjacent to the flowing section which leads to $d\varepsilon_w{}'=0$ as illustrated in Figure 3 [6]. The localising volume only deforms in thickness ($d\varepsilon_t$) and lengthways ($d\varepsilon_l{}'$), which characterises a biaxial strain state. Provided that there is an ideal plastic material behaviour and a negligible elastic fraction compared with the plastic fraction, Lévy-von Mises equations can be applied.

$$\frac{d\varepsilon_l}{\sigma_1 - \frac{1}{2} \bullet (\sigma_2 + \sigma_3)} = \frac{d\varepsilon_w}{\sigma_2 - \frac{1}{2} \bullet (\sigma_3 + \sigma_1)} = \frac{d\varepsilon_t}{\sigma_3 - \frac{1}{2} \bullet (\sigma_1 + \sigma_2)} \tag{1}$$

let $\sigma_1 \neq 0$ and $\sigma_2 = \sigma_3 = 0$ then

$$\frac{d\varepsilon_l}{\sigma_1} = \frac{d\varepsilon_w}{-\frac{1}{2} \bullet \sigma_1} = \frac{d\varepsilon_t}{-\frac{1}{2} \bullet \sigma_1} \tag{2}$$

If only $d\varepsilon_l$ and $d\varepsilon_w$ are considered (because of a small sheet thickness), the expression can be rewritten as

$$d\varepsilon_l = -2 \bullet d\varepsilon_w \tag{3}$$

Since $d\varepsilon_w < 0$ a Mohr´s circle of strain can be developed as illustrated in Figure 3. Applying a coordinate transformation it can be shown that α=70°32´if $d\varepsilon_w{}'=0$.

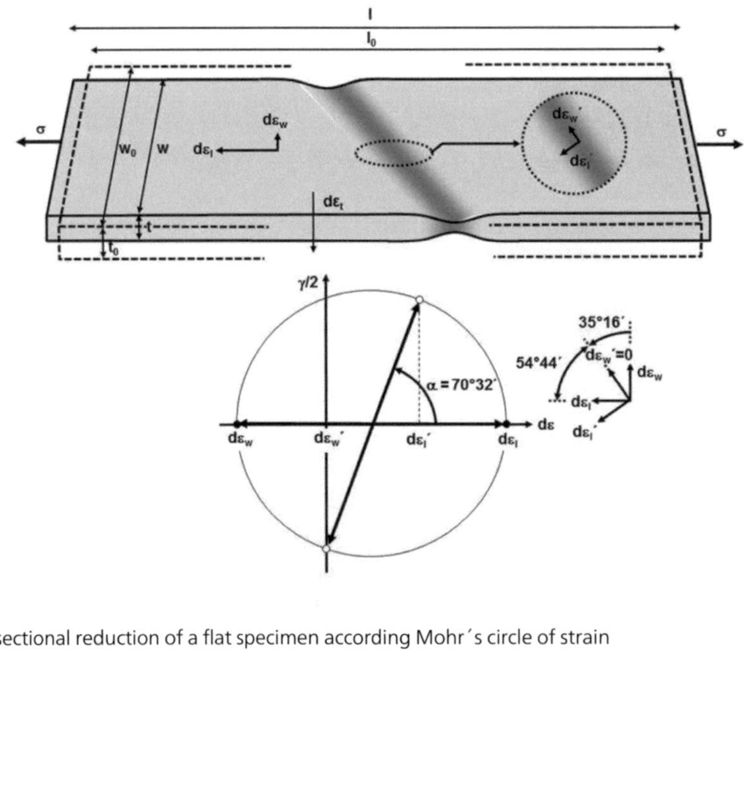

Figure 3. Cross-sectional reduction of a flat specimen according Mohr´s circle of strain

http://dx.doi.org/10.5772/58362

Thus, the angle between the direction of the localised neck and the direction of the principal normal stress is ideally 54°44′ which is in accordance with the hypothesis of Hill [7]. Eventual variations can be attributed to textures of the material.

2.2. Microscopic background of the PLC-effect

The microscopic background of the PLC effect is still a subject of discussion. This is because the PLC effect appears at temperatures where interstitial or substitutional diffusion of solute alloying elements do not offer sufficient mobility by bulk diffusion. A brief introduction of the development of ideas concerning the PLC effect is illustrated in [8] [9]. The first model developed by Cottrell and Bilby [10] assumed that the mobility of solute atoms is actually high enough to follow mobile dislocations. Subsequently, McCormick recognised that the mobility of solute atoms is barely sufficient to follow mobile dislocations, and he postulated that clustering mobile dislocations only occurs if they are arrested by obstacles such as by forest dislocations, for example [11]. Van den Beukel presupposed in his model, which was based on McCormick's work, that arresting mobile dislocations at forest dislocations is only caused by volume diffusion of solute atoms [12]. Following a proposal of Sleeswyk [13], the model created by Mulford [14] not only postulates the arresting of the stopped dislocation by volume diffusion, but additionally by a simultaneously running pipe diffusion (see Figure 4).

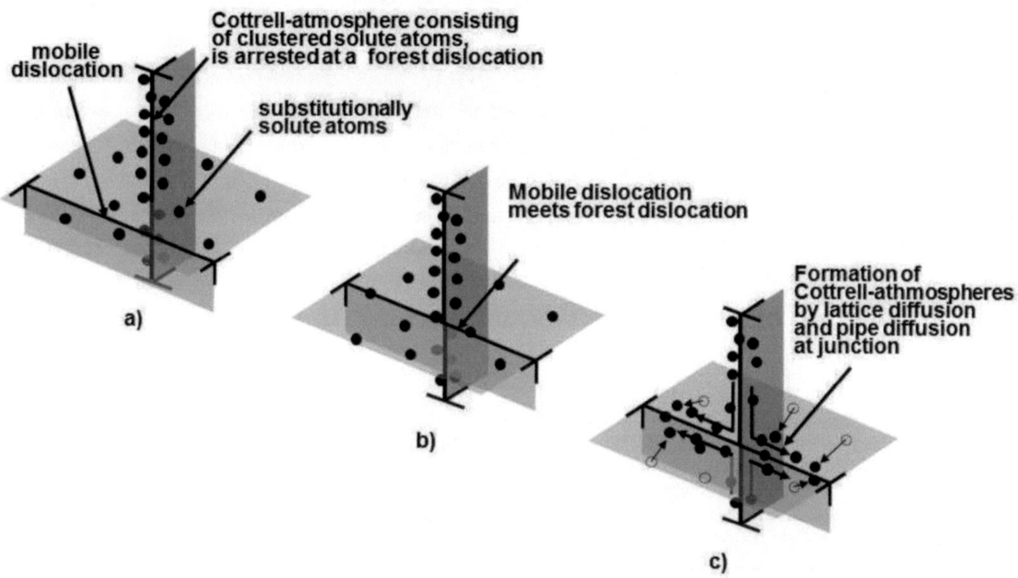

Figure 4. Substitutional solute atoms at stationary forest dislocations generate Cottrell-atmospheres, whose atoms cluster at a mobile dislocation which meets a forest dislocation via pipe diffusion.

A requirement for such a condition is a prior formation of a Cottrell-atmosphere at stationary forest dislocations. The pipe diffusion occurs without the acceleration caused by the vacancy mechanism. Modelling the pipe diffusion of the substitutionally dissolved magnesium in the aluminium bulk showed that the effectiveness of this diffusion process is not high enough to explain the experimental results [9]. As a consequence of this model the diffusion velocity of magnesium atoms is too slow to form Cottrell-atmospheres around the core of the arrested dislocation during representative times. Considering these results Picu [8] developed a new model which is based on the estimation that a forest dislocation embedded in a Cottrell-atmosphere reacts with a mobile dislocation by forming *dislocation jogs*. The higher the intensity of Cottrell-atmosphere around a forest dislocation the greater the energy necessary for forming jogs.

3. Experimental methods

A set of experimental tests were conducted based around a tensile test machine. Four types of wrought aluminium alloys were examined which are detailed in Table 1. In this table the sheet thickness and detailed composition are shown.

	EN AW-1050A	**EN AW-5182**	**EN AW-6016**	**EN AW-7021**
Condition	H14	H111	W_{30}– T4	
Temper [15]	H14 strain-hardened to a half hard condition - no thermal treatment	H111 slightly strain-hardened (less than H11) – no thermal treatment	T4 solution heat-treated at 465°C for 20 minutes and naturally aged to a substantially stable condition W30 solution heat-treated at 465°C for 20 minutes and tested within 30 minutes	
Specimen thickness [mm]	0,95	1,15	1,15	0,95
Si [wt.-%]	0,25	0,20	1,20	0,07
Fe [wt.-%]	0,40	0,35	0,40	0,16
Cu [wt.-%]	0,05	0,15	0,15	0,12
Mn [wt.-%]	0,05	0,35	0,15	0,02
Mg [wt.-%]	0,05	4,40	0,40	-
Cr [wt.-%]	-	0,10	0,10	-
Zn [wt.-%]	0,07	0,25	0,15	6,24
Ti [wt.-%]	0,05	0,10	0,15	0,02
Zr [wt.-%]	-	-	-	0,14

Table 1. Heat treatment, thickness and chemical composition of tested aluminium alloys

http://dx.doi.org/10.5772/58362

For the purpose of evaluating potential strain rate dependencies the uniaxial tensile tests for the aluminium alloys EN AW-6016 and EN AW-7021 in the solid solution heat treated and naturally aged condition T4, and the EN AW-5182 aluminium alloy in the strain hardened condition H111 (as currently implemented in industrial applications), were carried out with the following three forming velocities; $v_1=200mm/min$ (maximum velocity of the tensile testing machine), $v_2\left(\frac{1}{4}v_1\right)=50mm/min$ and $v_3\left(\frac{1}{10}v_2\right)=5mm/min$ resulting in strain rates $\dot{\varepsilon}_1(v_1)=6.60\times10^{-2}\ s^{-1}$, $\dot{\varepsilon}_2(v_2)=1.70\times10^{-2}\ s^{-1}$ and $\dot{\varepsilon}_3(v_3)=0.17\times10^{-2}\ s^{-1}$ respectively (see Table 2). By way of comparison, the highest strain rate illustrated, v_1 , approaches the forming velocity used in a typical industrial process. Determining the effect of precipitations for the 6000 and 7000 series a number of subsequent tests with the highest strain rate $\dot{\varepsilon}_1$ were carried out within 30 minutes after a solid solution heat treatment (W_{30} condition). As a reference the aluminium alloy EN AW-1050A was tested in half-hard condition (H14) with $\dot{\varepsilon}_1$. To be thorough regarding a potential influence of the strain rates, the reference aluminium alloy was also tested with the lowest strain rate ($\dot{\varepsilon}_3$)

Aluminium alloy	condition	strain rates		
		$6.60\cdot10^{-2}\ s^{-1}$	$1.70\cdot10^{-2}\ s^{-1}$	$0.17\cdot10^{-2}\ s^{-1}$
EN AW-1050A	H14	X		X
EN AW-5182	H111	X	X	X
EN AW-6016	W_{30}	X		
	T4	X	X	X
EN AW-7021	W_{30}	X		
	T4	X	X	X

Table 2. Different strain rates used in the experiment

Flat sheet specimens were CNC machined showing a test section width of 12.5mm in accordance with DIN EN ISO 6892-1 (A_{50}) [16]. Attention was given towards the rolling direction of the sheet metal which was perpendicular to the applied load. The length of the 50mm wide clamp sections of the specimens were designed to fit with the requirements of the cryogenic set-up.

A specially designed cryogenic dewar (Figure 5) was used to obtain and maintain the sub-zero temperatures required in these tests. A variation of testing temperatures was reached by way of a *medium-pipe*, which was used as a vessel carrying isothermal media, and was cooled by liquid nitrogen. The isothermal media used for the experiment shown in Table 3 were first pre-cooled in an external chamber. The final temperature was reached in the cryogenic dewar, where the specimen remained for around 5 minutes, before the test started. Furthermore, homogeneity of the temperature within the medium-pipe was accomplished by injecting inert gas. In case of a sudden temperature drop the chamber included an inlet for compressed air, which ensured a replacement of the cold environment between medium-pipe and cryogenic dewar. The investigations were conducted using a type Z100 tensile testing machine from Zwick.

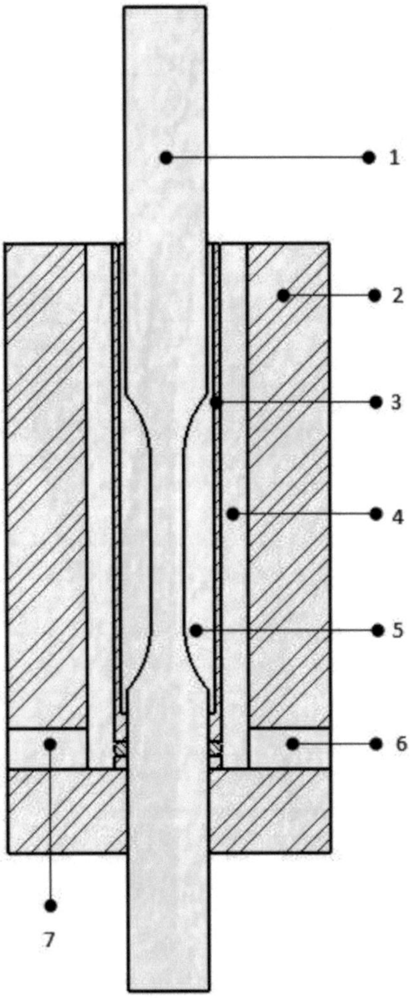

1) Specimen
2) Cryogenic dewar
3) Medium-pipe
4) Nitrogen filling
5) Medium filling
6) Gas inlet
7) Air inlet

Figure 5. Cryogenic set-up

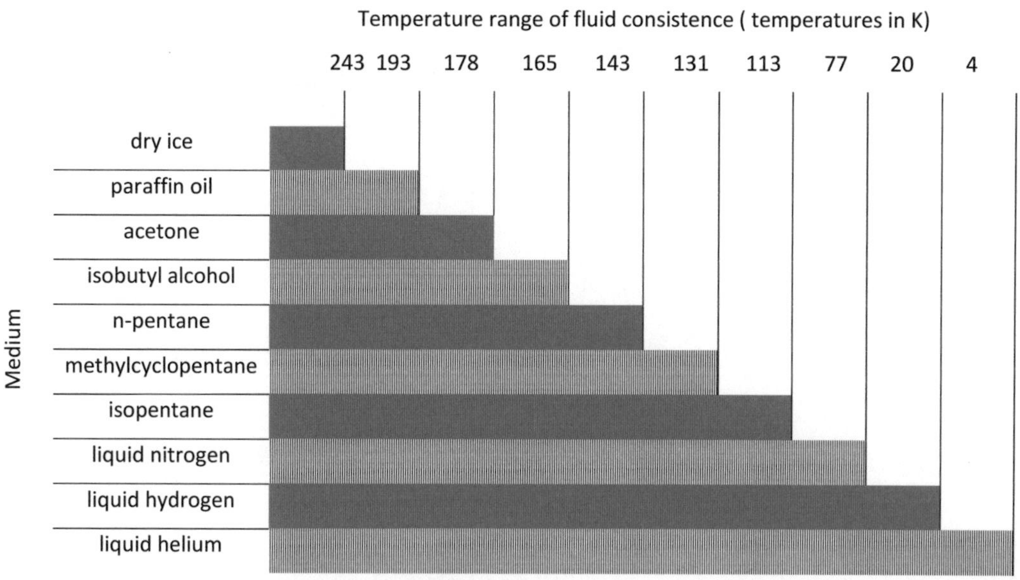

Table 3. Application temperature for different media [17]

4. Results

Figure 6 a) through f) shows for the aluminium alloys tested the temperature dependency of the yield strength σ_y, ultimate tensile strength (UTS), uniform elongation e_u and fracture elongation e_f which was determined at a strain rate of 6.60×10^{-2} s^{-1}.

It can be seen that with one exception and for all materials the yield strength, ultimate tensile strength and elongation rises consistently when temperature is decreased. The strength values of the aluminium alloys in W_{30} condition are lower than in T4 condition, which is as expected. For the aluminium alloys EN AW-5182 and EN AW-7021 in W_{30} condition a uniform elongation of over 40% and a fracture elongation of considerably more than 50% can be reached at 77K. A similar trend can be observed for the aluminium alloy EN AW-6016 in both heat treatment conditions, where the elongation in T4 condition is less than in W_{30} condition. For a decrease in temperature the aluminium alloy EN AW-1050A shows a rather shallow rise of the ultimate and yield strengths over the whole temperature range; whereas curves of uniform and fracture percentage elongation of this alloy indicate significant increases in the magnitudes of their gradients. The strength and elongation of the aluminium alloys EN AW-7021 also show a growing trend if temperature is reduced. For both heat treatment conditions, T4 and W_{30}, a gradual elongation rise of 25% at 152 K can be seen. If the temperature is lower than 152 K the elongation of the peak aged aluminium (T4) alloy decreases again. In turn, when present in W_{30} condition, the uniform elongation rises to more than 40% and the fracture elongation approaches 50% at 77K.

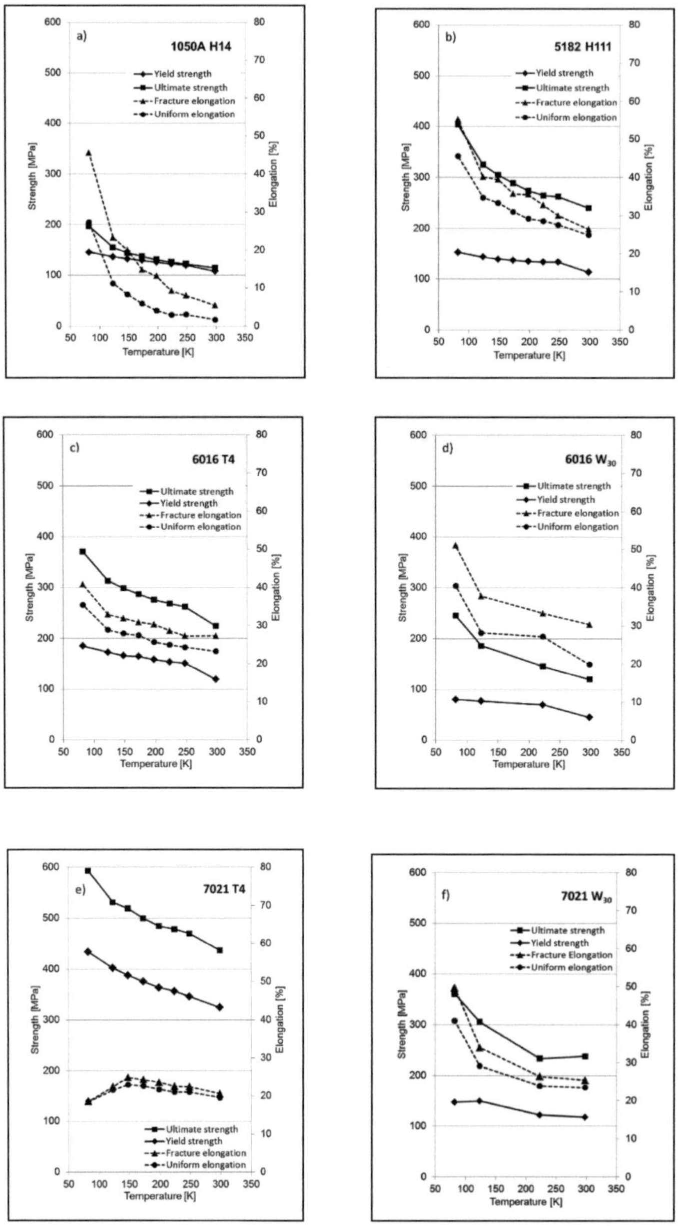

Figure 6. Temperature dependence of the yield strength σ_y, ultimate tensile strength UTS, uniform elongation e_u and fracture elongation e_f which was determined at a strain rate of $6.60 \cdot 10^{-2}\ s^{-1}$.

The influence of strain rates regarding strength values can be considered as small (see Figure 7 a), c), e) & g)). A slightly more distinct effect can be observed for the elongation values, especially for EN AW-5182 and EN AW-6016 T4. Nevertheless, those variations converge as the temperature approaches 77K (see Figure 7 b), d), f)&g).

http://dx.doi.org/10.5772/58362

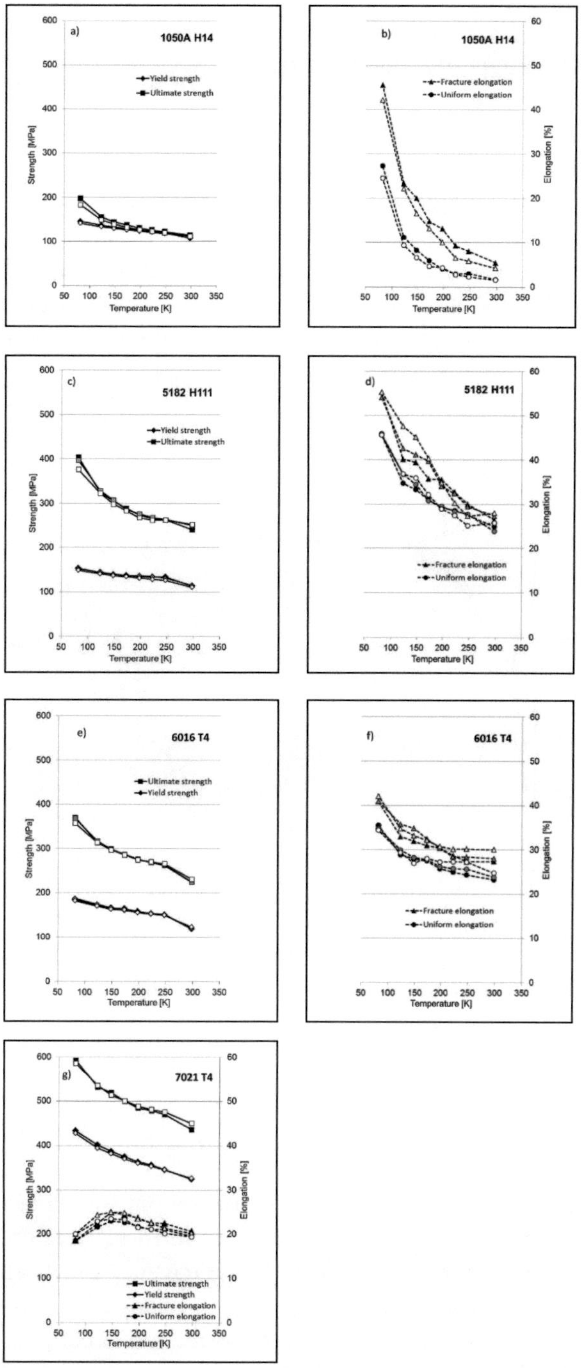

Figure 7. Temperature dependency of yield strength, ultimate strength and fracture elongation at different strain rates (black icon 6.60×10^{-2} s^{-1}, grey icon 1.70×10^{-2} s^{-1}, white icon 0.17×10^{-2} s^{-1})

Some overall observations where made as a result of the tests from stress-strain data. Firstly aluminium alloy EN AW-1050A-H14 showed no pronounced yielding for all strain rates and temperatures. The same effect was observed for EN AW-6016 in both heat treatment conditions, T4 and W_{30}. In contrast the aluminium alloy EN AW-5182-H111 showed pronounced yielding at all strain rates and temperatures. Aluminium alloy EN AW-7021 in W_{30} condition indicated pronounced yielding only at 298K and 248K. When the temperature was reduced this effect was not evident. On the other hand, EN AW-7021 in T4 condition showed similar yielding effects for all strain rates as was also observed for EN AW-6016. If one now considers Figure 8, the results of increasing the uniaxial load on an EN AW-5182-H111 specimen can be seen over the elastic and plastic regions. It is clear that there is a pronounced yielding behaviour accompanied by a region of stress oscillations. For the results, shown at T=298K, the amplitude is as high as ± 2 MPa; whereas for T ≤ 248K the magnitude reduces to ± 0.2 MPa.

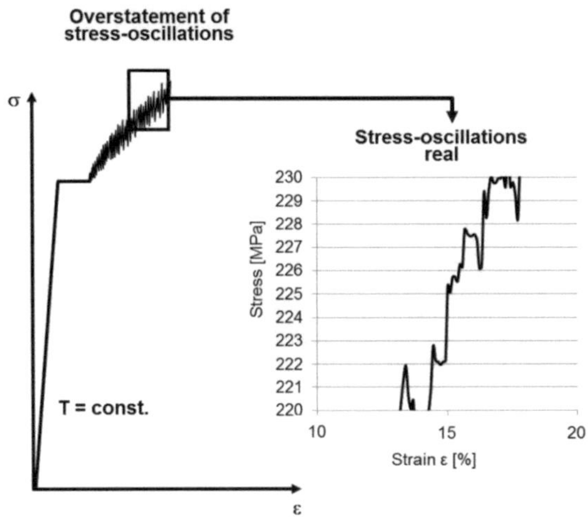

Figure 8. Pronounced yielding and stress amplitudes in the plastic region of EN AW-5182

5. Discussion

The PLC-effect, caused by Cottrell-atmospheres, offers a plausible mechanism to explain the resulting reduction of fracture elongation e_f and uniform elongation e_u, which leads to a decrease in the strain hardening coefficient n. The intensity of Cottrell-atmospheres is proportional to the quantity of dissolved atoms with a sufficiently high diffusivity, and the time to form Cottrell-atmospheres. Thus, in the case of the precipitation hardenable alloy EN AW-6016 and EN AW-7021, it was intended to increase the amount of dissolved atoms via solution heat-treatment, quenching and immediate testing of the mechanical properties at this condition. Comparing the two heat-treatment conditions T4 and W_{30}, it can be seen in Figure 6 c) through f) that this procedure had the expected positive effect on fracture elongation e_f

http://dx.doi.org/10.5772/58362

and uniform elongation e_u. It seems that the dissolved atoms had not the opportunity to form Cottrell-atmospheres which confirms the assertion of [8]. If the temperature is reduced, the diffusivity of the dissolved atoms is generally decelerated. Therefore, the formation of Cottrell-atmospheres, of the same intensity during the same period of time, is unlikely: this would correspond to a reduced PLC-effect. Figure 7 b), d) &f) illustrates this effect in terms of a growth in the fracture elongation e_f and uniform elongation e_u; thereby, the strain hardening coefficient n rises simultaneously as illustrated in Figure 9 a), b), c), d) & f). The overall increase in strength throughout the tested aluminium alloys could be explained *inter alia* by the activation energy which reduces as temperature is lowered. Consequently, the force needed to move a dislocation through the crystal lattice (Peierls Nabarro force), reaches a maximum as temperature converges to 0 K [4] [18].

Ludwik, Hollomon and Backofen identified that the flow stress K_f can be elevated by raising the strain ε as well as lifting the strain rate $\dot{\varepsilon}$ [19]. The general relationship between the physical values is:

$$K_f = K_f(\varepsilon,\ \dot{\varepsilon}) \tag{4}$$

A relationship which was empirically developed by Ludwik modified by Hollomon and extended by Backofen is

$$K_f = c \bullet \varepsilon^n \bullet \dot{\varepsilon}^m \tag{5}$$

where c is a material dependent constant (strength coefficient), n being the strain-hardening coefficient and m represents the strain rate sensitivity parameter. At temperatures where a metallurgical softening takes place to a limited extent ($T \leq 0.4T_m$; T_m=melting point [K]), an increase of the strain rate will not affect the true stress. Therefore, the m value can be assumed to be much smaller than 1. The results in Figure 7 confirm this statement.

If m is set to be 0 it follows:

$$K_f = c \bullet \varepsilon^n \tag{6}$$

Using this relationship most stress-strain curves can be described accurately in the region of uniform elongation at $T \leq 0.4T_m$. As exemplified in Heiser [5] the strain hardening coefficient n can be written as true uniform strain ε_u or related to the uniform elongation e_u.

$$n = \varepsilon_u = ln\left(1 + e_u\right) \tag{7}$$

The strain hardening coefficient n describes to what extent a material can be deformed before localised necking will start, i.e. the flow curve. Figure 9 shows correlations between experimental evaluated strain hardening coefficients using Eq. (7) and uniform elongation e_u in relation to temperature and strain rate of EN AW-1050A, EN AW-5182, EN AW-6016 and EN

AW-7021 specimens. In addition, the Ludwik curve is included which represents n calculated using Eq. (7).

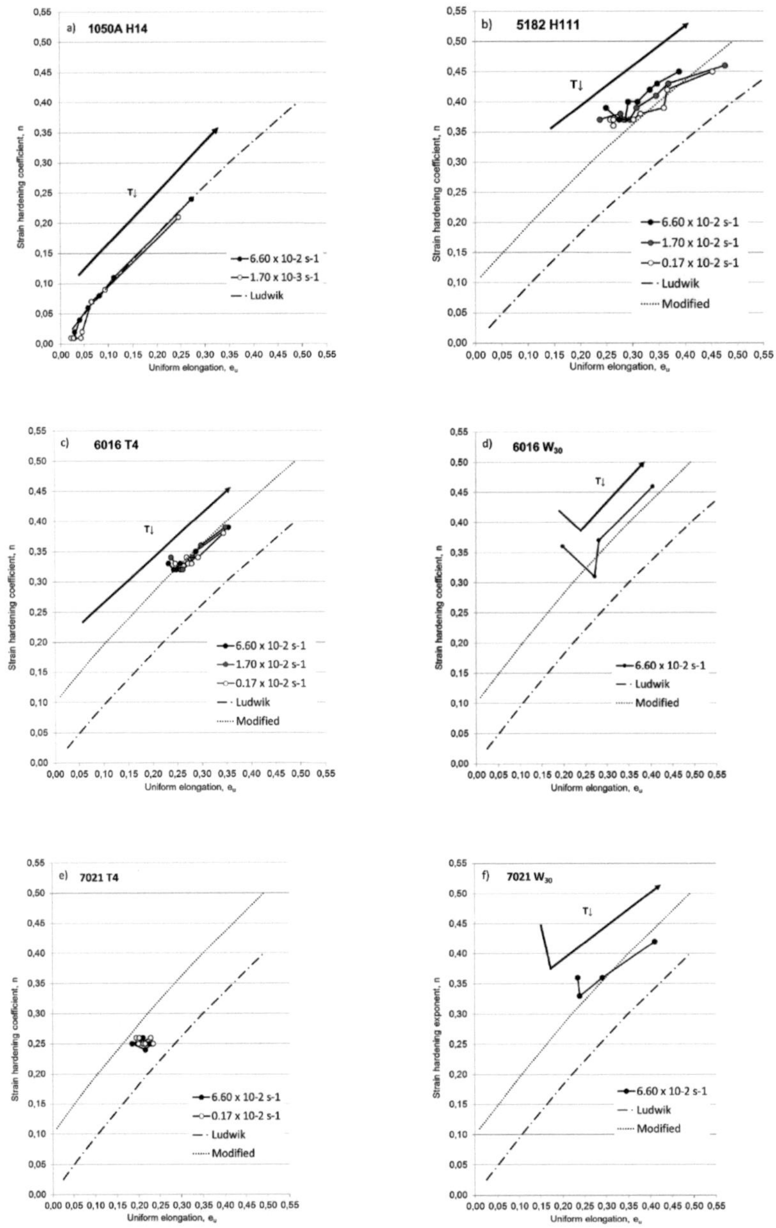

Figure 9. Temperature and strain rate dependence of strain hardening exponent n and uniform strain ε_u. Additionally included is the dependence of strain hardening coefficient and uniform elongation after Ludwik if only strain hardening is considered

Figure 9 a) through f) shows only a slight strain rate dependency. Apart from the temperature range between 298K and 248K and from EN AW-7021 in the precipitation hardened condition (T4), an increase of the strain hardening coefficient and uniform elongation as temperature is reduced can be seen. Except for the aluminium alloy 1050A, (Figure 9 a)) which is consistent with the Ludwik curve, tested aluminium alloys indicate a parallel offset. It is assumed that this shift is due to the PLC-effect. Therefore Eq. 6, which takes only the strain hardening process into account, seems not to be applicable for all aluminium alloys. A modification to this equation would seem appropriate to allow for the description of materials which exhibit a PLC-effect. Eq. (8) illustrates a *simple* modification of Eq. (7). The addition of a constant, p, provides a shift to the Ludwik curve and can now be written

$$n = ln\ (1 + e_u) + p \tag{8}$$

The modified Ludwik curve is illustrated on Figure 9 b) through f) which incorporates a value of p=0.1. Whilst it is clear that the new fit corresponds more appropriately (recalling that the 7021 Figure 9 e) is in a naturally aged condition (T4) and is not in an ideal state for cryogenic forming processes), further analyses of flow curves presented in this and additional studies will provide the basis of a more refined relationship. This will be presented in a subsequent publication.

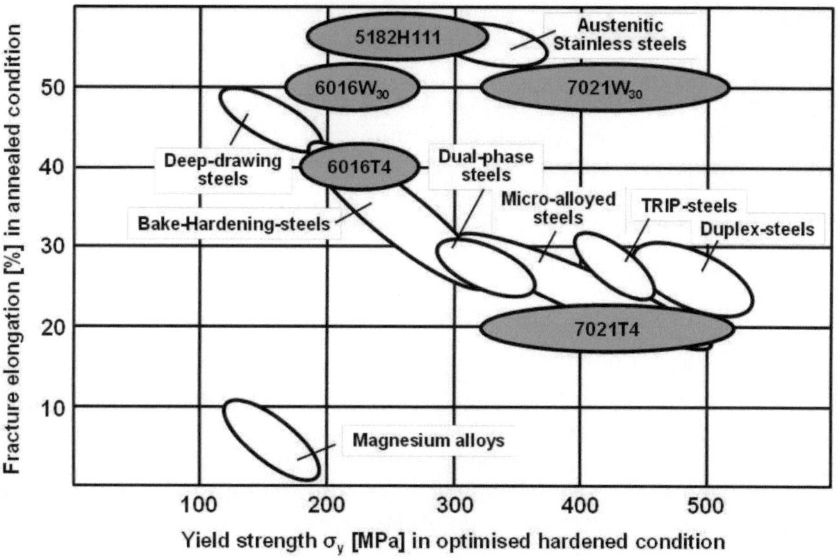

Figure 10. Fracture elongation for aluminium wrought alloys of the EN AW-5xxx, EN AW-6xxx and EN AW-7xxx series at T=77 K and deep-drawing steels at T=298 K. Yield strength in optimised hardened condition for all mentioned groups at T=298 K.

6. Conclusion

The plastic behaviour of commercial aluminium wrought alloys in annealed or solution annealed condition can be positively enhanced if the forming temperature is below 123K. This advantage can be reached without any interference of higher strain rates.

Figure 10 demonstrates the fracture elongation values which are currently attainable from aluminium alloys compared with values achievable from deep-drawing steels. A competitive advantage of cryogenic forming of annealed or solution annealed aluminium alloys is clearly shown. Flow curves showing PLC-effects can be described more precisely with addition of a constant p which represents an extension of the initial Ludwik relationship.

Author details

R. Schneider[1,2], B. Heine[2] and R.J. Grant[1]

1 Glyndŵr University, Department of Engineering and Applied Physics, United Kingdom

2 Aalen University of Applied Sciences, Institut für Oberflächen-und Werkstofftechnik, Germany

References

[1] W. S. Miller, L. Zhuang, J. Bottema, A. J. Wittebrood, P. De Smet, A. Haszler and A. Vieregge, "Recent development in aluminium alloys for the automotive industry," Mat. Sci. Eng. A280 (2000) 37-49.

[2] E. Pink and A. Grinberg, "Praktische Aspekte des Portevin-LeChatelier-Effektes (1)," Aluminium 60 (1984) 687-691.

[3] E. Pink and A. Grinberg, "Praktische Aspekte des Portevin-LeChatelier-Effektes (2)," Aluminium 60 (1984) 764-768.

[4] G. E. Dieter, Mechanical Metallurgy, London: McGraw-Hill BookCompany (UK) Limited, 1988 pp. 289-291.

[5] M. Heiser and G. Lange, "Scherbruch in Aluminium-Legierungen infolge lokaler plastischer Instabilität," Z. Metallkd. 83 (1992) 115.

[6] S. P. Keeler and W. A. Backofen, "Plastic instability and fracture in sheets stretched over rigid punches," Trans. ASM 56 (1963) 25-48.

[7] R. Hill, "On discontinuous plastic states, with special reference to localized necking in thin sheets," J. Mech. Phys. Sol. 1 (1952) 19-30.

[8] R. C. Picu, "A mechanism for the negative strain-rate sensitivity of dilute solid solutions," Acta Mat. 52 (2004) 3447-3458.

[9] R. C. Picu, G. Vincze, F. Ozturk, J. J. Gracio, F. Barlat and A. M. Maniatty, "Strain rate sensitivity of the commercial aluminum alloy AA5182-O," Mat. Sci. Eng. A390 (2005) 334-343.

[10] A. H. Cottrell and B. A. Bilby, "Dislocation theory of yielding and strain ageing of iron," Proc. Phys. Soc. A62 (1949) 49-62.

[11] P. G. Mc Cormick, "A model for the Portevin-Le Chatelier effect in substitutional alloys," Acta Metall 20 (1972) 351-354.

[12] A. van den Beukel, "Theory of the effect of dynamic strain ageing on mechanical properties," Status Solidi A30 (1975) 197-206.

[13] A. W. Sleeswyk, "Slow strain hardening of ingot iron," Acta Metall 6 (1958) 598-603.

[14] R. A. Mulford and U. F. Kocks, "New observations on the mechanisms of dynamic strain aging and of jerky flow," Acta Metall 27 (1979) 1125-1134.

[15] DIN EN 515: Aluminium und Aluminiumlegierungen-Halbzeug-Bezeichnung der Werkstoffzustände, Berlin: Beuth Verlag, 1993.

[16] DIN EN ISO 6892-1: Zugversuch-Metallische Werkstoffe-Prüfverfahren bei Raumtemperatur, Berlin: Beuth Verlag, 2009.

[17] "IFA Institut für Arbeitsschutz der Deutschen Gesetzlichen Unfallversicherung," GESTIS Substance Database, [Online]. Available: http://gestis-en.itrust.de [Accessed 10 02 2014].

[18] B. Heine, Werkstoffprüfung Ermittlung von Werkstoffeigenschaften, München: Carl Hanser Verlag, 2011 pp.170-172.

[19] J. K. Gregory, Superplastic deformation in oxide dispersion strengthened nickel base superalloys, Stanford University: Ph.D. Thesis, 1983.

Chapter 4

Recrystallization Behavior of Commercial Purity Aluminium Alloys

Rajat K. Roy

Additional information is available at the end of the chapter

http://dx.doi.org/10.5772/58385

1. Introduction

Commercial purity aluminium alloys are largely used in the forms of foil for food packaging industries. Aluminium ingots are processed through the multi-operational steps in the consequence of foil preparation. The ingot of 7mm thickness, obtained from either direct chill (DC) casting or continuous casting process, is first cold rolled to reduce the thickness approximately 0.6 mm followed by an intermediate annealing treatment and then final roll pass is given to produce the foils of desired thickness. Therefore, the intermediate annealing treatment and consequent recrystallization behavior are prime factors for controlling microstructures as well as properties of foils. In general, the recrystallization behavior is the resultant of three steps, viz., deformation, recrystallization and grain growth. It is affected by various impurities and the precipitates forming at the operational steps.

Iron and silicon are the inevitable impurities in commercial purity aluminium alloys. The precipitation reaction due to these impurities easily occurs during heat treatment at the temperature range of 200 – 600°C, affecting cold working, softening, and corrosion resistance of the alloys. The type of intermetallic phases formed during solidification and volume fraction, amount and size of the individual particles can be controlled by changing the silicon (Si) and iron (Fe) content, i.e., by adjusting the Fe/Si ratio [1]. The silicon influences the nature of eutectic constituent, solid solubility of other elements, formation of precipitates and dispersoids, and the transformation characteristics of the precipitates [2]. It is necessary to control the Si content of high quality aluminium foils where material characteristics are determined by Fe/Si ratio. If the material is properly processed, then it can only be assumed that the material properties are not actually deteriorated with increased amount of Fe, Si and/or Mn. In general, composition ratio of the various alloying elements is maintained with respect to each other, but the absolute value of the single alloying element is not critical as long as primary crystallization of coarse intermetallic phases is avoided.

2. Plastic deformation

2.1. Microstructure of deformed metals

The cold-rolled alloys are consisting high density of dislocations clearly examined through transmission electron microscopy (TEM). During deformation the dislocations are arranged as cell boundaries whereas some cell interiors are free of dislocations (Fig. 1). Due to heavy deformation, the cell boundaries are diffuse in nature instead of simple dislocation.

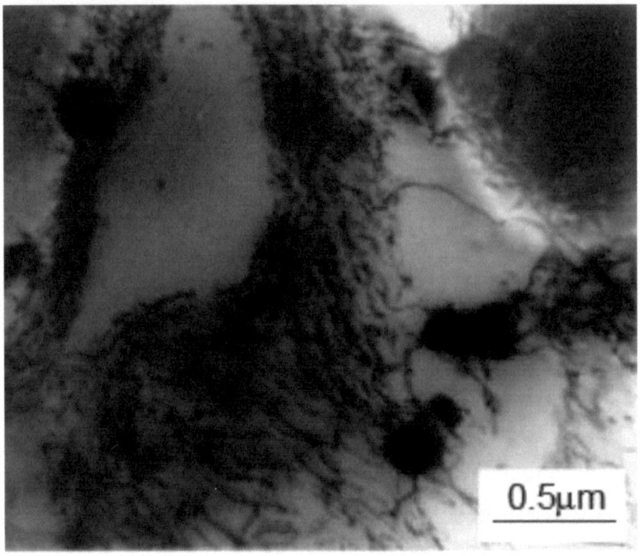

Figure 1. Dislocation structures of 92% cold rolled commercial purity aluminium alloy (AA1235 :0.67 Fe, 0.16Si, 0.01 Ti and rest Al) [3]

However, the orientation of plastically deformed alloy changes from grain to grain or in different regions in a same grain. It happens owing to different rotations during the deformation by use of a different combination of slip systems to achieve the imposed strain. Therefore, the deformation band is nothing but a volume of constant orientation that is significantly different than the orientation of other parts of the grain, which is explained by a schematic diagram (Fig. 2). It shows the variation of orientation at different regions of a same grain. The orientation of one part (X) of a deformed grain changes rapidly to that of a differently oriented part (Y) of the same grain across a thin boundary of a finite width called the transition bands (T) or microbands. Sometimes the orientation of deformation bands is changed twice, such as from X to Y and then Y to X. This special type of deformation band is called kink band.

http://dx.doi.org/10.5772/58385

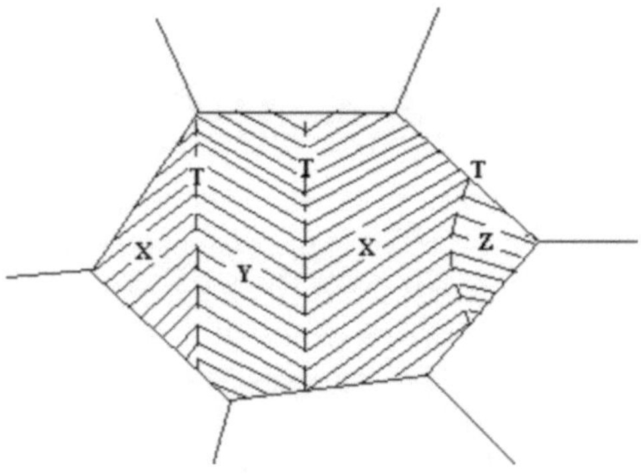

Figure 2. Schematic diagram of deformation bands, transition bands and kink bands [4]

After plastic deformation by slip mechanism, the walls of high dislocation density separate the region of low dislocation density. Such types of microstructure have been referred to as cell or subgrain. In the case of cell structures, the boundary consists of a tangled array of dislocations and appears to be a diffuse in nature. For subgrain, the boundary is sharp and consists of a well ordered dislocation array. The cell structure develops in the bulk material during or after the dislocation movement induced by the applied stress. With increasing strain, the walls become quite sharp and interiors become reasonably free of dislocations. At this stage the cell might be called subgrain.

2.2. The effect of particles on deformed microstructure

Due to rapid cooling rate in the continuous casting and direct chill casting, the alloying elements remain as supersaturated solid solution. Upon the processing operation of aluminium foils, the increased cold rolling reduction gives rise to a more pronounced dislocation entanglement. It results in the acceleration of both stored energy and precipitation. Therefore, during plastic deformation of two-phase alloys, particles affect the overall dislocation density, inhomogeneity of deformation in the matrix and deformation structure. Subsequently, the recrystallization behaviour of the alloy is also affected due to the influence of driving force as well as nucleation sites for recrystallization. If the particle is strong enough to withstand the applied stress on it, the dislocation then proceeds to encircle the particle and leaves an Orowan loop, otherwise it deforms. When the particle does not deform then extra dislocations are generated at the particle-matrix interface. On the other hand, if the particle deforms either before or after the formation of Orowan loop, no extra dislocations are generated. Generally, smaller particles are weaker than larger particles, which results in weaker slip plane. Movement of subsequent dislocations occurs on the same plane. As a consequence, slip bands are formed.

2.3. Deformation texture

The cold rolling texture of aluminium alloy sheets has been characterized as β fiber, which is associated with plane strain deformation [5]. The texture is varied from layer to layer of the sheet. The surface layer of the rolled sheet is known to have the shear texture, which is different from that of the plane strain deformed center layer [6-10]. According to Grewen and Huber, the ideal orientations in 95% cold rolled aluminium are characterized by {112}<111>, {110}<112> and {123}<412> [11]. The penetration depth of shear texture increased with increasing friction coefficients and rolling temperature [7]. Above all, material parameters also influence the shear texture formation. The most important factors among material parameters are the yield strength and strain hardening exponent [8,9]. The texture inhomogeneity has often been of considerable importance in rolling of pure aluminium, which has a low yield strength and work hardening exponent and tends to develop the shear texture in the surface layer [12-14]. Aluminium and copper single crystals of the S-orientations develop cube texture after cold-rolling by more than 97.5% and subsequent annealing [15, 16]. This occurs owing to the close correlation between the formation of {001}<100> deformation structure and the development of cube texture. The cube-oriented material formed during rolling is considered to be the preferential nucleation sites of cubically aligned recrystallized grains. The factors which affect the generation of the {001}<100> deformation structure and the equivalent components during rolling are process parameters like friction between roll and sheet, rolling temperature, lubricants and the L/d value (where L and d represent the contact length and the thickness of specimens, respectively). Low viscosity lubricant such as kerosene suppresses the formation of {001}<100> deformation structure, whereas using machine oil as lubricant a cube component is formed. Thereafter, subsequent annealing treatment develops retained rolling and random texture in kerosene treated alloy and sharper cube texture in machine oil treated alloy [17]. Truszkowski et al. [9, 10] and Asbeck and Mecking [8] have reported that the shear texture is developed when $L/d \geq 5$ or $L/d \leq 0.5$.

3. Recovery

3.1. Subgrain formation and its growth

During plastic deformation of polycrystalline material (specially the alloy of medium and high stacking fault energy), unequal numbers of dislocations of two signs are generated, and the dislocations are rearranged as three-dimensional cell structure with complex dislocation tangles as cell walls. At recovery stage, the dislocations of opposite sign annihilate each other by combination of gliding and climbing mechanisms. The excess dislocations are left in the material at the end of the first stage of recovery. Upon progressing stages, these excess dislocations are arranged in a low energy configuration in the form of regular arrays or low angle grain boundaries (LAGBs). This mechanism is called polygonization [18], and the newly formed cells are called subgrains. In other words, the dislocation movement from the cell interiors to the cell boundary causes low angle grain boundaries, resulting in the subgrain formation. Sometimes, the dynamic recovery also helps to form a well-developed subgrain structure during deformation. In this case, recovery is the only involvement for the coarsening of subgrain structure. Although low stacking fault energy material shows poor subgrain structure, but when fine second-phase particles inhibit recrystallization then at high temperature a well-defined substructure is formed by recovery [19]. Coarsening of substructure takes place by two methods-subgrain boundary migration and subgrain rotation.

The cell structure of a heavily cold rolled alloy is modified as a function of annealing time, examining the subgrain growth. In a commercial purity aluminium alloy (AA1235), after annealing at 250ºC for 1 h, the dislocations are examined to rearrange themselves to form a subgrain structure with an average diameter of 0.60 μm. Dislocation networks form the low angle boundary also known as subboundary. The dislocation rearrangement in the regions where the cell structure is not well-developed is described as a "disentanglement of dislocations" (Fig. 3a). Consequently, the dislocation density decreases inside the subgrain and particles are precipitated at subboundary as well as inside of subgrains (Fig. 3b). After 4 h of annealing at same temperature homogeneous subgrain growth takes place with an average diameter of 2 μm. Generally, the sharp delineation of many subboundaries decreases upon annealing and ultimately fades away, in agreement with the proposed coalescence mechanism (Figs. 3c). At this stage, subgrain growth is hindered by the pinning action of Al_3Fe precipitates.

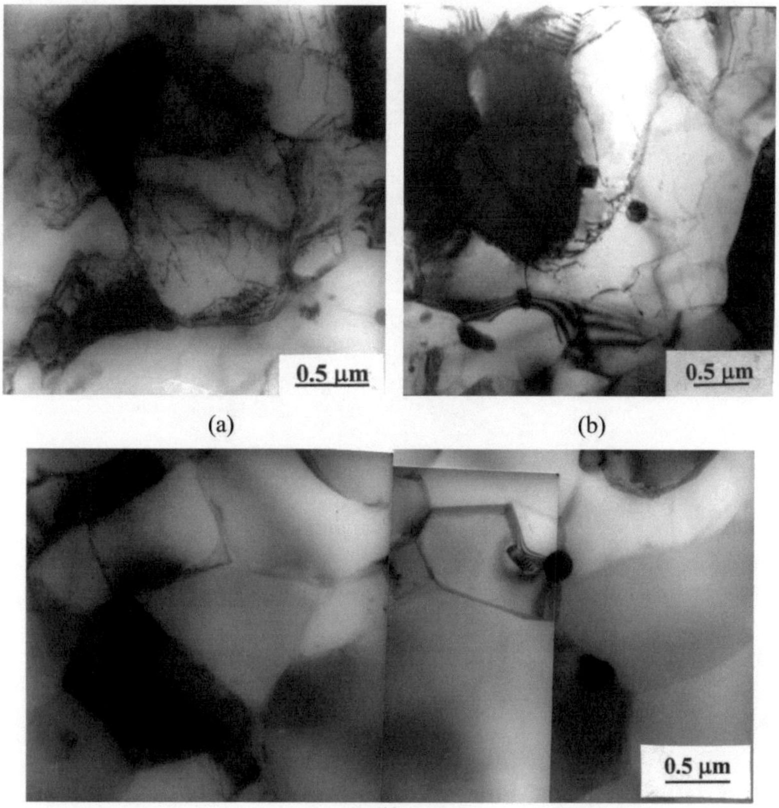

(c)

Figure 3. TEM micrographs (BF) showing subgrain formation of 92% cold rolled alloy AA1235 annealed at 250ºC for (a) & (b) 1h and (c) 4h.

3.2. The effect of second phase particles

In two phase alloys, second phase particles inhibit recovery by the pinning of individual dislocation or dislocations of the low angle boundaries. Therefore, the dispersion of fine particles exerts a strong pinning effect on the subgrains [20-22]. Moreover, the particles stabled at high temperature improve the strength and creep properties of the alloy by pinning and stabilizing the recovered substructure. If the particles prevent the subgrain to reach the critical size for the formation of recrystallization nucleus, then recrystallization is also hindered. The evidence of dislocation pinning and restriction of subgrain growth is examined for a commercial purity aluminium alloy (AA8011) after annealing below 400ºC [23]. Similarly, the well-defined subgrains are examined in another commercial purity aluminium alloy (AA1235) after 4h annealing at 250ºC [24], as shown in Fig.4. The dislocation density change is clearly examined at some portions (X) of boundaries. The distribution of dislocation is also another observation at some points (Y). At this stage, dislocations accumulate at the grain boundary region and sometime they result in the formation of grain boundary lines at Z [25]. It can be said that the effect of particles on subgrain growth is similar to the effect of particles on grain growth.

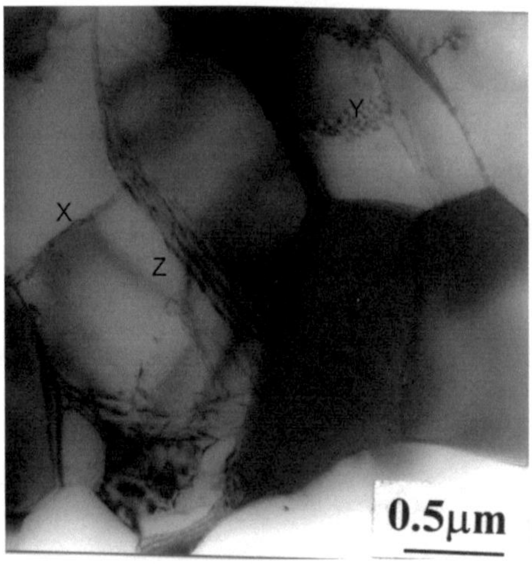

Figure 4. TEM micrographs (BF) showing the effect of annealing at 250°C for 4 h on subgrain structure of 92% cold rolled bulk specimen (AA1235).

4. Recrystallization

4.1. The effect of deformation percentage

Nucleation and growth of new recrystallized grains are closely dependent on the distribution of dislocations within the deformed grains, relating to the deformation percentages. Recrystallization does not happen below a minimum strain and the start temperature of recrystalli-

http://dx.doi.org/10.5772/58385

zation decreases with increasing the minimum strain level. It is known that at low strain (<5%) high activation energy for nucleation results in a low nucleation rate and a relatively high growth rate [26]. With varying cold rolling percentage for different commercial purity aluminium alloys (AA1145, AA1200, AA8011), the degree of softening is identical for low reduction levels (<95%) and less steep for reduction levels above 95% (Fig.5) [27]. It corresponds to the gradual coarsening of microstructure with increasing temperature for high reduction levels (>95%), relating to the continuous recrystallization.

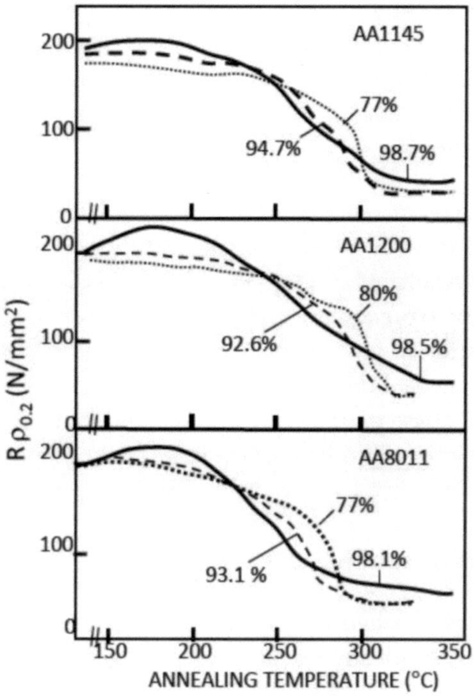

Figure 5. Yield stress variation with annealing temperature for three commercial purity aluminium alloys at different percentage of cold rolling [27]

4.2. Nucleation of recrystallization

The recrystallization nuclei originate at preferred sites such as prior grain boundaries, transition bands, and shear bands of deformed microstructure. The nucleation process at the grain boundary region by two different mechanisms:

1. **Strain induced boundary migration,** where a nucleus forms when an original grain boundary bulges out [28].

2. **Subgrain coarsening by coalescence at the original grain boundary,** where a nucleus forms in the consequence of subgrain growth [29-31].

With an increasing degree of deformation, the number of potential nucleation sites such as deformation bands and grain boundary bands increases significantly and the number of such sites is larger in fine grain than in coarse grain specimens [32]. The effectiveness of nucleation sites is enhanced by the presence of intermetallic particles (Al_3Fe) with increasing degree of deformation. Potential nucleation sites are more important factor than number of nuclei. If the same numbers of nuclei form at the grain boundaries of both fine grain material and coarse grain material, then the fine grain material shows more homogeneous recrystallization. In case of heterogeneous recrystallization, all grains are not recrystallized at the same rate. It is known that crystallographic orientation affects the slip systems and strain path during deformation. Therefore, distribution and density of dislocations, large scale microstructural inhomogeneties, availability of nucleation sites, and growth rate of recrystallized grains are also dependent on crystallographic orientation.

4.3. Sequence of precipitation and recrystallization

It is well established that the recrystallization behavior of a deformed and supersaturated alloy is largely dependent on whether or not precipitation of the second phase particles take place simultaneously with recrystallization [33]. Hornbogen and Köster have suggested that recrystallization occurs prior to precipitation at high temperatures, whereas precipitation takes place prior to recrystallization at low temperatures [34]. If particles are precipitated during recrystallization they may hinder both the formation and migration of recrystallization fronts. Alternatively, if the second phase particles are precipitated in the matrix prior to cold rolling, the recrystallization behaviour will depend on the size and dispersion of the second phase particles.

4.4. The effect of inter particle spacing

At wide inter particle spacing, when only a few particles are present, there is basically no difference between the recrystallization behaviour of two-phase and single phase alloys. In this case, nucleation generally occurs at the original matrix boundaries and the ultimate recrystallized grain size may vary owing to inhomogeneous distribution of nucleation sites. With an increase in particle content of the alloy (i.e., decrease in interparticle spacing) nucleation occurs more rapidly at the lattice curvature of particle-matrix (for the particles >1 μm) interface than at the grain boundary region [35]. In addition, due to the increase in quantity and uniform distribution of particle-matrix nucleation sites, the final recrystallized grains become more uniform and finer. However, the trend towards increased nucleation due to decrease in interparticle spacing occurs until particles maintain a critical spacing (C_1) to allow the nucleation to occur simultaneously and independently at each particle. After reaching the critical spacing (C_1), a nucleus attached to one particle would be viable itself, if it left to develop by itself. It would be non-viable if nuclei start to form simultaneously at neighbouring particles. They would then interfere with each other before reaching a viable size. Thus, due to the formation of fewer amount of viable nuclei at particle-matrix interfaces during increase in particle content (i.e., decrease in interparticle spacing) the overall nucleation rate of recrystallization decreases. Further increase in particle content leads to a second critical spacing (C_2) at which viable nuclei formation is inhibited owing to the proximity of particles. In such circumstances nucleation is likely to occur predominantly at the original matrix grain boundaries. As a consequence, the nucleation as well as recrystallization rate drastically get reduced if the inter particle spacing falls below C_2 [36].

4.5. The effect of particle size

At the time of deformation, large particles (> 1 μm) lead to a heterogeneous distribution of dislocations, whereas fine particles (< 0.1 μm) give rise to a homogeneous distribution of dislocations. As a consequence, the number of possible recrystallization nucleation sites increases for large particles and decreases for fine particles. This can be related to the degree of deformation (Fig. 6), explaining the nucleation of recrystallized grains at the particles greater than 2 μm for a highly deformed metal [37]. The fine particles also inhibit sub-boundary migration and thus the nucleation process is retarded. As the particles become more finely spaced, nearly all cell boundaries will be pinned by the particles at the end of the deformation stage. In this case, when recrystallization occurs, the mechanism of nucleation is not well understood, but the kinetics of the process is certainly very slow [38]. Closely spaced, thermally stable particles preserve the deformed/ recovered microstructure up to the melting point of the matrix. Kim et al. have revealed that the retention of dislocation substructure at high temperature provides an additional strengthening mechanism to the dispersion hardening of the alloys used at high temperature structural applications [39]. With the decrease in the particle size and spacing between particles, recrystallization kinetics is retarded, but final grain size becomes quite large [37, 40]. Several researchers have studied the relationship between dispersion characteristics, deformation substructure, and recrystallization [41-43].

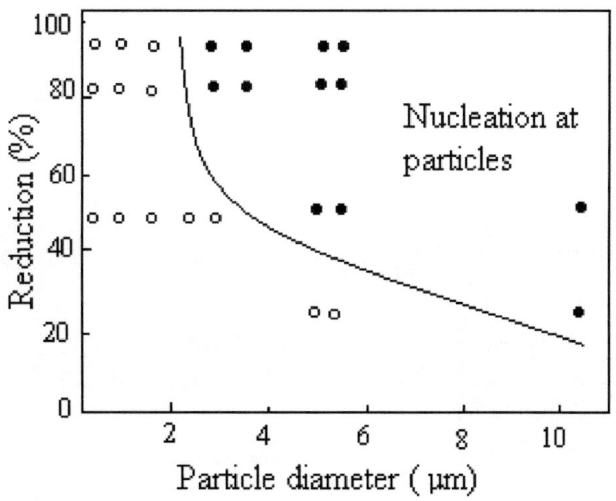

Figure 6. The conditions of deformation and particle size for which nucleation of recrystallized grains is observed to occur at particles [37].

4.6. Type of particles in commercial purity aluminium

In commercial purity aluminium alloys, the main impurity Fe combines with both Al and Si to form a large variety of phases during solidification or during subsequent thermomechanical processing [44, 45]. Shoji and Fujikura have identified three types of precipitates Si, α-AlFeSi, and Al_3Fe in cold rolled commercial alloy AlFe0.6Si0.16 (all in wt%) after annealing in salt bath at different temperatures 300, 400 and 500ºC, respectively [3, 46]. Some researchers have identified only Al_3Fe precipitates during recrystallization of commercial Al-Fe-Si alloy [47]. In AA8011 alloy, Al_3Fe and Si precipitates are observed in cold rolled and annealed conditions [48]. After heavy plastic deformation (~92%), very small particles of 0.17 μm size are distributed in the dislocations [3]. Upon progress of annealing, the particle size increases, and both smaller and larger, spherical and plate-shaped Al_3Fe particles are inhomogeneously distributed and situated at subboundaries and subgrain interiors. Since the dislocations cannot cross the grain boundary, during annealing a large variation of dislocation density occurs from grain to grain owing to the pinning effect of precipitate particles. After completion of recrystallization, particle also inhibits the grain boundary migration along with pinning of the dislocations. Even after a long time of high temperature annealing for heavily deformed alloy, the Al_3Fe particles may pin down the dislocations, resulting in the presence of a large amount dislocations inside the grains (Fig. 7) [3, 24]. During the intersection of grain boundary by particles, a Zener drag force is generated to restrain the boundary migration [49]. The coherent particle generally loses the coherency when a high angle grain boundary moves past a coherent particle. The coherent particles are twice as effective in pinning a grain boundary as incoherent particles of the same size [4]. There are many alternative situations during particle-boundary interaction. The particle may dissolve during passage of the boundary and re-precipitate in a coherent orientation, it may reorient itself to a coherent orientation, or the boundary may cut through the particle [50].

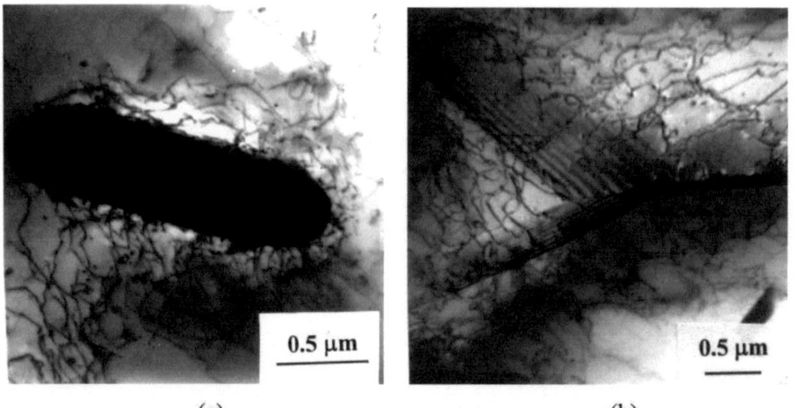

(a) (b)

Figure 7. TEM micrographs (BF) of 92% cold rolled alloy AA1235 after annealing at 480ºC for 8 h showing (a) pinning of dislocations by a particle, and (b) presence of dislocations even after completion of recrystallization

http://dx.doi.org/10.5772/58385

4.7. Particle distributions in bimodal alloys

Many commercial alloys are bimodal alloys which contain both large (>1 μm) and small particles. Large particles act as nucleation sites for recrystallization and small particles hinder the grain boundary migration, i.e., retard the recrystallization. Therefore, recrystallization behaviour of bimodal alloys is affected by particle distribution. During deformation, the fine dispersion of small particles does not affect significantly the degree of lattice curvature generated at the large particle-matrix interface. Later, by altering the dispersion parameters of particles, recrystallization kinetics and microstructures are controlled. Chan and Humphreys have reported that in the bimodal alloy, nucleation of recrystallization takes place at the large particles, but fine particles determine the time for completion of recrystallization [51]. They have observed a large number of small island grains and coarse irregular grains in the microstructure of bimodal alloys. With increasing coarsening of the fine precipitates, the grain shapes become more regular and the mean size decreases. Nes has taken the help of the parameter *f*/*r* (where *f* and *r* are volume fraction and radius of small particles, respectively) to describe a model to account for the grain size of bimodal alloys [52]. In bimodal alloys, critical particle size for the growth of a nucleus is

$$d = \frac{4\gamma}{P_d - P_z} = \frac{4\gamma}{\frac{\rho G b^2}{2} - \frac{3f\gamma}{2r}} \tag{1}$$

where P_d is driving force (stored energy in the subgrain boundaries),

ρ is dislocation density,

G is shear modulus,

b is burgers vector,

P_z is Zener pinning force,

is grain boundary energy,

f is volume fraction of small particles,

r is radius of small particles

As *f*/*r* of fine dispersion increases, the large particles act as the nucleation sites, but growth of the grains is slowed down at the early stage of annealing. However, the earlier developed grains either consume the smaller grains, or form island grains.

4.8. Recrystallization texture

Annealing of heavily deformed material leads to a wide range of recrystallization textures. Different types of recrystallization textures can be produced in similar alloys each having almost identical and very strong deformation textures [11]. In f.c.c. metals, sometimes recrystallization texture contains cube texture along with retained rolling texture and in some cases nearly random textures [11]. It is reported that with increasing deformation, the volume fraction of the cube texture decreases preferably for low annealing temperatures and that of the rolling texture increases for all temperatures, whereas the random part of the texture strongly decreases [53]. The possible origin of recrystallization texture has been reviewed several times [54, 55]. It is notable that "oriented nucleation" and "oriented growth" are two possible theories for the development of recrystallization texture [56, 57]. The recrystallization texture develops from the competition between cube oriented grains which nucleate at the cube bands, R-oriented grains (i.e., retained rolling texture) which stem from the grain boundaries between the former deformed grains, and randomly oriented grains due to particle stimulated nucleation (PSN).

4.8.1. *The effect of particles*

It is already discussed in previous section (§ 4.4) that widely spaced coarse particles (>1 μm) enhance nucleation and the rate of recrystallization if it is present before deformation. The finely dispersed particles normally reduce recrystallization kinetics having a greater retarding influence on nucleation than on growth. Therefore, the texture changes can be rationalized on the basis of the orientation dependent nucleation. If nucleation of all components is retarded in equal proportions, then the time available for growth of the first formed grains is increased. The recrystallization texture will therefore show increased selectivity and will become even more strongly biased towards the most favoured orientation nuclei. The effect of particle size on the formation of major components of recrystallization texture is shown in Table 1 [58].

Material (% of reduction)	Particles		Recrystallization Texture
	Volume fraction (f)	Diameter (μm)	
Al (99.9965) (90%)	0	-	Cube
Fe-AlN (70%)	0.06	0.017	Rolling
Al-Al_2O_3 (30-90%)	0.4	0.1	Deformation
Cu-SiO_2 (70%)	0.5	0.23	Rolling + Twins
Al-FeSi (90%)	0.5	0.2- 7	Cube, Rolling, Random
Al-Si (90%)	0.8	2	Cube, Rolling, Random
Al- Ni (80%)	10.0	1	Rolling, Random
Al- SiC (70%	20.0	10	Random

Table 1. Major components of the recrystallization texture for some particle containing alloys [58]

The effect of PSN on recrystallization texture is as follows [59]:

1. PSN results in a sharp recrystallization texture for low strained material or in case of particle containing single crystals. When PSN originates in several deformation zones at

a single particle of a lightly deformed polycrystals, a spread of orientation is observed around the particles in the deformed matrix.

2. In the case of heavily deformed material, the presence of different deformation zones of different grains or different deformation regions of the same grain results in a single grain with a wide range of orientations. As a consequence, either a weak texture or randomly oriented grains form.

The recrystallization texture is dependent on the particle size, strength and spacing. When the particles precipitate during annealing, the recrystallization texture developed is similar to the deformation texture. There is no clear reason for the retention of the rolling texture in the recrystallized alloys. It appears that a strong cube texture is rarely formed and particles may be responsible for suppressing the formation or viability of cube sites. This enables other components including the retained rolling components to dominate the texture.

Although PSN nuclei are oriented randomly in the heavily rolled polycrystal, the final texture of this alloy is not random. Therefore, it is concluded that grains from other sites with different orientation affect the final texture. From Table 2.2 it is observed that for the alloys (e.g., Al-Fe-Si or Al-Si) containing low volume fraction of particles, cube and rolling components are developed, whereas rolling and random components develop for high volume fraction of particles containing alloys (e.g., Al-Ni). The cube grains are larger than randomly oriented grains owing to the faster growth rate of former [59]. It results in formation of small island grains inside the large cube grains. Some researchers have shown that the strength of the cube component decreases with an increase in the number of supercritical sized particles in the case of hot-rolled aluminium alloy AA3004 [60]. On the other hand, volume fraction of randomly oriented grains increases with an increase in the number of supercritical sized particles. Hornbogen et al. have distinguished between 'discontinuous' (conventional recrystallization) and 'continuous' recrystallization [61]. The former cases give rise to normal recrystallization textures, such as the cube texture, while the latter cause retention of the rolling texture.

4.8.2. Effect of Fe on the commercial purity aluminium

A typical recrystallization texture of highest purity aluminium is the cube-orientation {001}<100> [5, 11]. In commercial purity aluminium the iron and silicon contents have been found to be important factors in controlling the recrystallization texture [62, 63]. It is reported that a small amount of Fe may cause a change almost from pure cube to retained rolling texture [4]. During recrystallization, precipitation of Al_3Fe particles inhibits the growth of the early formed cube nuclei, thereby forcing nucleation in the abundant rolling texture components. Thus, the recrystallization textures of most commercial aluminium alloys are composed of three texture types, namely the cube texture, the retained rolling texture and the random texture [64, 65]. Cube-oriented subgrains are known to rapidly recover either dynamically during deformation or statically during the early stage of recrystallization, which gives rise to a size advantage of cube nuclei [66]. Growth of the dominant cube grains is lowered by solute drag and/or precipitation, which leads to the development of rolling component and reduces the strength of the cube component. In a sheet, a mixture of cube and rolling components is desired to prohibit earing during deep drawing of that sheet [54]. It is interesting to note from the research work of Hirsch and Lücke that 95% cold-rolled alloy Al-0.007% Fe shows the retained rolling texture with a small amount of cube texture at 360ºC compared to strong cube texture at 280 and 520ºC [67]. It is due to the formation of iron-rich phase at 360ºC, which restricts cube grain growth. When precipitation occurs before or after completion of recrystallization, the effect on boundary migration is less drastic and a strong cube texture develops. It has been observed that the growing grains of cube component can consume the recrystallized grains of non-cube component along with the surrounding deformed material [68]. It is reported that coarse iron rich particles inhibit the formation of cube texture in cold rolled aluminium alloys [11]. It is evident that increasing iron content decreases the strength of the cube texture, especially when it is present in solid solution form before rolling [11]. Even the addition of 0.1% iron to aluminium may change the recrystallization texture [69].

4.8.3. Effect of recrystallization texture on microstructure and mechanical properties

Grain size and mechanical properties can be controlled by maintaining proper texture in recrystallized microstructure. It is already discussed in §4.8.2 that Al_3Fe particles suppress the cube texture formation. Cube grain growth is also inhibited by "orientation pinning" [59]. Orientation pinning occurs when a recrystallized grain of a given orientation grows into deformed material of its own orientation. It results in a low angle boundary and the growth of the recrystallized grain will practically stop due to the low mobility of low angle boundaries. The case where orientation pinning is expected to be important is for the growth of grains with the same orientation as a main component in the deformation texture, like retained rolling components. Vatne et al. have also reported a case where the growth of cube grains has been most likely reduced due to orientation pinning. This may be due to an extremely high as-deformed cube fraction of 35% [70]. Therefore, fine grains can be generated in the foil by controlling recrystallization texture during intermediate annealing treatment, which improves the mechanical properties of the foil. Blade has demonstrated that earing tendency of aluminium alloy increases with increase in cube texture component [71]. It is shown that earing tendency becomes zero when 25% cube texture is present.

5. Grain growth

5.1. The effect of second-phase particles and orientation gradient

During grain growth, boundary migration is retarded by a Zener drag effect of the second-phase particles. As the driving pressure of grain growth is extremely low, particles may have a very large influence on both the kinetics of grain growth and the resultant microstructure. Humphreys et al. have elaborately discussed the growth related formulation, i.e., driving and pinning pressures for growth, grain growth rate, limiting grain size due to particles, etc. [4]. In the case of planar grain boundary, Zener limiting grain size (D_{Zener}) which equals to the critical radius of island grains, arises from the balance between the driving pressure for grain growth (P) and Zener pinning pressure (P_z).

$$D_{Zener} = \frac{4r}{3f} \tag{2}$$

Equation 2 is only applicable to those materials where the second-phase particles are stable during grain growth. Therefore, stability of the particles with temperature is one of the important factors for growth mechanism. The instability of the second-phase particles, i.e., precipitation after the formation of grain and subgrain structure or coarsening of particles during grain growth, affects the grain growth by a different way and it has been discussed elsewhere [4].

Orientation gradient has a greater effect on the grain boundary mobility. High stacking fault energy alloys (Al, Ni and α-iron etc.) readily form a cellular or subgrain structure during deformation. This structure is often not uniform and orientation gradients are usually developed here. In addition, many alloys produced by thermomechanical processing develop a preferred orientation or texture by recrystallization and grain growth. A model developed by Ferry illustrates that the grain coarsening is rapid for large orientation gradient and particle-free systems, and it is markedly reduced in a system containing a large volume fraction of fine particles, despite the presence of the orientation gradient [72]. The latter case occurs due to retardation of the onset of recrystallization.

5.2. The competition between normal and abnormal grain growth

Grain growth may be divided into two types, normal grain growth and abnormal grain growth. When the grains grow uniformly with a narrow range of size and shapes and the grain size distribution is independent of time, then, this type of grain growth is called "normal" or "continuous" grain growth. In "abnormal" or "discontinuous" grain growth, few grains are large compared to the rest and grain size distribution is bimodal. Since this discontinuous growth of selected grains has similar kinetics to primary recrystallization and has some microstructural similarities, abnormal grain growth is sometime called secondary recrystallization. The normal grain growth theory is based on the grain boundary interfacial free energy as the driving force. If the initial grain size distribution is too wide, a fraction of large grains will grow in an abnormal manner until all the other grains have been consumed. When completed, this process has resulted in a more narrow size distribution and at longer times the steady state may be approached asymptotically. It appears convenient to define this state as a normal grain growth. Abnormal grain growth may sometime be a necessary initial stage toward find normal grain growth [73].

In spite of considerable research efforts, the origin of abnormal grain growth is yet to be fully understood. However, by Monte Carlo simulation it has been observed that two different conditions may initiate abnormal grain growth [74]. Firstly, anisotropy in the grain boundary energy may lead to rapid growth of grains having boundary energies much lower than the average [75]. This is often the case when the material exhibits a strong primary recrystallization texture [76]. Secondly, abnormal grain growth may arise from anisotropy in the grain boundary mobility [74, 77].

Hillert has deduced three conditions for the development of abnormal grain growth in a material as follows [73]:

1. Normal grain growth cannot take place due to particle pinning,
2. The average grain size has a value below the limit 1/2z (where$z=3f/4r$, f and r are the volume fraction and the radius of particles, respectively).
3. There is at least one grain much larger than the average.

Whether these conditions are automatically fulfilled in a material where the normal grain growth has stopped growing due to the presence of second phase particles, is a question of considerable practical importance. Again, the growth of a very large grain in a material where the normal grain growth has stopped depends upon the value of the final grain size reached by the normal grain growth. The mathematical analysis, on the other hand, predicts that normal grain growth should proceed up to the limit 1/2z. This causes abnormal grain growth to be impossible for a limited time period when the normal grain growth has started to slow down but has not yet reached the limit. Therefore, initiation of abnormal grain growth is possible by a continuous decrease in the z value. This effect can be accomplished increasing the particle size r through coalescence or decreasing the volume fraction f by dissolving the second phase. Any process that leads to a slow increase in the grain size limit may initiate the development of abnormal grain growth. Most cases of abnormal grain growth, met within practice, seem to be connected with the dissolution of the second phase rather than the coalescence [73].

The occurrence of abnormal grain growth may be limited by "nucleation" rather than growth considerations. It is evident from a large number of publications that abonormal grain growth is likely to occur as the annealing temperature is raised and as the particle dispersion becomes unstable [76, 78-79]. If a single strong texture component is present in a fine-grained recrystallized material, then abnormal grain growth commonly occurs on further annealing at high temperatures [80, 81]. This is due to a low misorientation and hence low energy and mobility of grain boundaries in highly textured materials. Due to the presence of another texture component, higher energy and mobility are introduced in the boundaries to migrate preferentially by a process, which is closely related to primary recrystallization. The avoidance of abnormal grain growth at elevated temperatures is an important aspect of grain size control in steels and other alloys. A safe method of avoiding abnormal grain growth would be to form an average grain size so much larger than the particle-limited grain size [73]. Another method is by choosing a material with a larger volume fraction of second phase particles. In other words, it can be said that abnormal grain growth is not likely to occur where most of the particles are observed to be situated at the grain boundaries.

http://dx.doi.org/10.5772/58385

Acknowledgements

The author acknowledges helpful discussions with Prof. S. Das, Prof. K. Das, Department of Metallurgical and Materials Engineering, Indian Institute of Technology, Kaharagpur 721302, INDIA.

Author details

Rajat K. Roy

MST Division, CSIR-National Metallurgical Laboratory, Jamshedpur, India

References

[1] P. Furrrer, G. L. Schenks, and R.E. Upton, *Aluminium Alloys for Packaging,* Eds. By J. G. Morris, H. D. Merchant, E. J. Westerman and P. L. Morris (Warrendale, PA: The Minerals, Metals & Materials Society), (1993), p. 144.

[2] E. J. Westerman, *Aluminium Alloys for Packaging,* Eds. By J. G. Morris, H. D. Merchant, E. J. Westerman and P. L. Morris (Warrendale, PA: The Minerals, Metals & Materials Society), (1993), p. 3.

[3] R. K. Roy, S. Kar, K. Das and S. Das, *Mater. Let.,* Vol. 59 (2005), p. 2418.

[4] F. J. Humphreys and M. Hatherley, *Recrystallization and Related Annealing Phenomenon,* Published by Elsevier Science Ltd., (2002).

[5] J. Hirsch and K. Lücke, *Acta Metall.,* Vol. 36 (1988), p. 2863.

[6] G. Von Vargha and G. Wassermann, *Metallwirtschaft,* Vol. 12 (1933), p. 511.

[7] I. L. Dillamore and W. T. Roberts, *J. Inst. Met.,* Vol. 92 (1963-64), p. 193.

[8] H. O. Asbeck and H. Mecking, *Mater. Sc. Engg.,* Vol. 34 (1978), p. 111.

[9] W. Truszkowski, J. Krol and B. Major, *Metall. Trans. A,* Vol. 11A (1980), p. 749.

[10] W. Truszkowski, J. Krol and B. Major, *Metall. Trans. A,* Vol. 13A (1982), p. 665.

[11] J. Grewen and J. Huber, in : F. Hassener (Ed.), Recrystallization of metallic materials, Dr. Riederer Verlag GmbH, Stuttgart, 1978, p. 111.

[12] H. Hu and R. S. Cline, *Trans. TMS-AIME,* Vol. 224 (1962), p. 784.

[13] H. W. F. Heller, J. H. van Drop, G. Wolff and C. A. Verbraak, *Met. Sci.,* Vol. 15 (1981), p. 333.

[14] H. P. Kneijnserg, C. A. Verbraak and M. J. Ten Bouwhuijs, *Acta Metall.*, Vol. 33 (1985), pp. 1759-67.

[15] T. Kamijo, A. Fujiwara, Y. Yoneda and H. Fukutomi : *Acta Metall. Mater.*, Vol. 39 (1990), p. 1947.

[16] T. Kamijo, H. Adachihara, H. Fukutomi and E. Aernoudt, *Acta Metall. Mater.*, Vol. 40 (1992), p. 693.

[17] T. Kamijo, H. Morita, S. Kataoka and H. Fukutomi, *Materials Science Forum*, Vols. 113-115 (1993), p.145.

[18] F. J. Humphreys, in "*Processing of metals and alloys*", Ed. by R. W. Cahn, VCH, Germany, Vol. 9 (1991), p. 373.

[19] J. F. Humphreys and J. W. Martin, *Phil. Mag.*, Vol. 16 (1967), p. 927.

[20] J. F. Humphreys and J. W. Martin, *Phil. Mag.*, Vol. 17 (1968), p. 365.

[21] H. Ahlborn, E. Hornbogen and U. Köster, *J. Mats. Sci.*, Vol. 4 (1969), p. 94

[22] A. R. Jones, and N. Hansen, *Acta Metall.*, Vol. 29 (1979), p. 589.

[23] R. K. Roy, S. Kar, and S. Das, *J. Alloy & Comp.*, Vol. 468 (2009), p. 122.

[24] R. K. Roy, S. Kar, K. Das and S. Das, *J. of Mater. Sc.*, Vol. 41(2006), p. 1039

[25] H. Gleiter, *Acta Metall.*, Vol.17 (1969) p. 565.

[26] W. A. Anderson and R. F. Mehl, *Trans. Met. Soc. AIME*, Vol. 161 (1945), p.140.

[27] A. Oscarsson, H. E. Ekström, and B. Hutchinson, Mat. Sc. Forum, Vol. 113-115 (1993) pp. 177.

[28] P. A. Beck and P. R. Sperry, *J. Appl. Phys.*, Vol. 21 (1950), p. 150.

[29] N. Ryum, *Acta Metall.*, Vol.17 (1969) p. 269.

[30] R. D. Doherty and R. W. Cahn, *J. Less Com. Metals*, Vol. 28 (1972), p. 279.

[31] A. R. Jones, B. Ralph and N. Hansen, *Proc. R. Soc. Lond.*, Vol. A368 (1979), p. 345.

[32] B. Bay and N. Hansen, *Metall. Trans. A*, Vol. 15A (1984), p. 287.

[33] J. Holm and E. Hornbogen, *J. Mater. Sci.*, Vol. 5 (1970), p. 655.

[34] E.Hornbogen and U.Köster, in "*Rerystallization of Metallic Materials*", Ed. by F. Haessner, Reiderer-Verlag, Stuttgart, (1971), p. 159.

[35] P. R. Mould and P. R. Cotterill, *J. Mater. Sci*, Vol. 2 (1967), p. 241

[36] P. Cotterill and P. R. Mould, in "*Recrystallization and Grain Growth in Metals*", Surrey University Press, London, (1976).

[37] F. J. Humphreys, *Acta Metall.*, Vol. 25 (1977), p. 1323.

[38] M. A. Morris, M. Leboeuf and D. G. Morris, *Materials Science Forum*, Vols. 113-115 (1993) p. 257.

[39] Y-W Kim, W. M. Giffith, in "Dispersion Strengthened Aluminium Alloys",TMS, Warrendale, USA, (1988).

[40] R. D. Doherty and J. W. Martin, *J. Inst. Metals*, Vol. 91 (1962), p. 332.

[41] J. L. Brimhall, M. J. Klein and R. A. Huggins, *Acta Metall.*, Vol.14 (1966), p. 459.

[42] J. F. Humphreys and J. W. Martin, Bri 1966: J. L. Brimhall, M. J. Klein and R. A. Huggins, *Acta Metall.*, Vol.14 (1966), p.775.

[43] D. T. Gawne and G. T. Higgins, *J. Mater. Sci.*, Vol. 6 (1971), p. 403.

[44] D. Munson, *J. Inst. Met.*, Vol. 95 (1967), p. 217.

[45] A. L. Dons, *Z. Metallkd.*, Vol. 15 (1984), p. 170

[46] R. Shoji and C. Fujikura, *Key Engg. Mater.*, Vols. 44-45 (1990), p. 163.

[47] R. K. Davies, V. Randle and G. J. Marshall, *Acta Mater.*, Vol. 46 (1998), p. 6021.

[48] J. H. Ryu, D. N. Lee, *Mat. Sc. and Engg. A*, Vol. A336 (2002), p. 225.

[49] C. S. Smith, *Trans. Metall. Soc. AIME*, Vol. 175 (1948), p.15.

[50] R. D. Doherty, *Metal Sci*, Vol. 16 (1982), p. 1.

[51] H. M. Chan and F. J. Humphreys, *Acta Metall.*, Vol. 32 (1984), p. 235.

[52] E. Nes, *Proc. 1st Risø Int. Symp. on Met. Mat. Sci. Recrystallization and Grain Growth of Multi-Phase and Particle Containing Materials* (Ed. by N. Hansen, A.R. Jones and T. Leffers), (1980), p. 85.

[53] O. Engler, H.E. Vatne and E. Nes *Mat Sc. and Engg. A*, Vol. A 205 (1996), p. 187.

[54] W.B. Hutchinson, *Metal Science*, Vol. 8 (1974), p. 185.

[55] K. Lucke, *7th Int. Conf. on Textures of Materials*, The Netherland Society for Material Science, Zwijndrecht, Holland, (1984), p. 195.

[56] W. G. Burgers and P. C. Louerse, *Z. Physik*, Vol. 61 (1931), p. 605.

[57] C. S. Barrett, *Trans. Am. Soc. Metals*, Vol. 137 (1940), p. 128.

[58] F. J. Humphreys, D. Juul Jensen, *Proc. 7th Int. Risø Symp.*, Risø, Denmark, 1986, p. 93.

[59] H. E. Vatne, S. Benum, O. Daaland and E. Nes, *Textures and Microstructures*, Vols. 26-27 (1996) p. 385.

[60] R. K. Bolingbroke, G. J. Marshall R. A. Ricks, *Aluminium Alloys for Packaging*, Eds. By J. G. Morris, H. D. Merchant, E. J. Westerman and P. L. Morris (Warrendale, PA: The Minerals, Metals & Materials Society), (1993), p. 215.

[61] E. Hornbogen and H. Kreye, in *"Textures in Research and Practice"* Springer-Verlag, Berlin, (1969), p. 274.

[62] J. Grewen and M.V. Heimendahl, *Z. Metallk.*, Vol. 59 (1968), p. 205.

[63] D. H. Rogers and W. T. Roberts, *Z. Metallk.*, Vol. 65 (1974), p. 100.

[64] D. Juul Jensen, N. Hansen and F. J. Humphreys, *Acta Metall.*, Vol. 33 (1985), p. 2155.

[65] O. Engler, J. Hirsh and K. Lücke, *Acta Mater.*, Vol. 43 (1995), p. 121.

[66] A. L. Dons and E. Nes, *Mater. Sci. Tech.*, Vol. 2 (1986), p. 8.

[67] J. Hirsch and K. Lücke, *Acta Metall.*, Vol. 33 (1985), p. 1927.

[68] W. Bleck and H. J. Bunge, *Acta Metall.*, Vol.29 (1981) p. 1401.

[69] W. Bunk, *Z. Metallkunde*, Vol. 56 (1965), p. 645.

[70] H. E. Vatne and E. Nes, Proc. 16th Risø Int. Symposium, (Eds. N. Hansen et al.), Roskilde, Denmark, (1995), p. 581.

[71] J. C. Blade, *J. Aust. Inst. Metals*, Vol. 12 (1967), p. 55.

[72] M. Ferry, *Acta Mater.*, Vol. 53 (2005), p. 773.

[73] M. Hillert, *Acta Metall.*, Vol. 13 (1965), p. 227.

[74] D. J. Srolovitz, G. S. Grest and M. P. Anderson, *Acta Metall.*, Vol. 33 (1985), p. 2233.

[75] A. D. Rollet, D. J. Srolovitz and M. P. Anderson, *Acta Metall.*, Vol. 37 (1989), p. 1227.

[76] J.E. May and D. Turnbull, *Trans. Metall. Soc. AIME*, Vol. 212 (1958), p. 769.

[77] S. ling and M.P. Anderson, *JOM*, Vol. 44 (1992), p. 30.

[78] T. Gladman, *Proc. R. Soc. Lond.*, A294 (1966), p. 298.

[79] T. Gladman, in "*Grain growth in Polycrystalline Materials*" Ed. by Abbruzzese and Brozzo, Trans Tech Publns., Rome, (1992), p. 113.

[80] P. A. Beck and H. Hu, Trans. Metall. Soc. AIME, Vol. 194 (1952), p. 83.

[81] C. G. Dunn and P. K. Koh, Trans. Metall. Soc. AIME, 206, p. 1017.

Chapter 5

Machining Burrs Formation & Deburring of Aluminium Alloys

Seyed Ali Niknam, Yasser Zedan and Victor Songmene

Additional information is available at the end of the chapter

http://dx.doi.org/10.5772/58361

1. Introduction

Although the machinability of most aluminium alloys can be classified as relatively easy when the tool wear and the cutting energy are considered, these materials could however raise some concerns when the chip formation and the burr formation are of concern. Burr formation, a phenomenon similar to chip generation, is a common problem that occurs in several industrial sectors, such as the aerospace and automobile sectors. It has also been among the most troublesome impediments to high productivity and automation, and largely affects the machined part quality. To ensure competitiveness, precise and burr-free components with tight tolerances and better surface finish are demanded. Intensive research conducted during the last decades has laid out the mechanisms of burr formation and deburring in a very comprehensive fashion, and has introduced integrated strategies for burr prevention and minimization. Despite all the improvements realized, there are still many challenges encountered in understanding, modeling and optimizing the burr formation process and size, through production growth and cycle time reduction. Furthermore, acquiring a solid knowledge on deburring methods and the links between them and burr size is strongly recommended.

This chapter reviews burrs formation and the factors governing them, including the workpiece material, the tooling, the machining parameters and the machinining strategy. A case study on the effect of heat treatments on drilling burr size is also presented and is followed by some deburring and edge finishing techniques commonly used for machined aluminum parts. The main advantages, disadvantages and limitations of these edge finishing operations are also presented.

2. Overview of burr formation

Burr formation is one of the major issues currently facing manufacturing industries. During the process of plastic deformation, the material is stretched past the point of elastic deformation, where it can no longer return to its original shape and size. If there is already a crack present in the material, the stretching will continue to increase the size of the crack eventually causing it to fracture [1]. Therefore, burrs forming during machining are defined as projections of material beyond the workpiece limits [2]. It is very important to limit burr formation rather than deburring them in a subsequent finishing operation [3]. Burr consists of an undesirable extended surface over the workpiece [4] or a missing portion on the workpiece edge (negative burr, see Figure 2), which should be avoided or at least minimized.

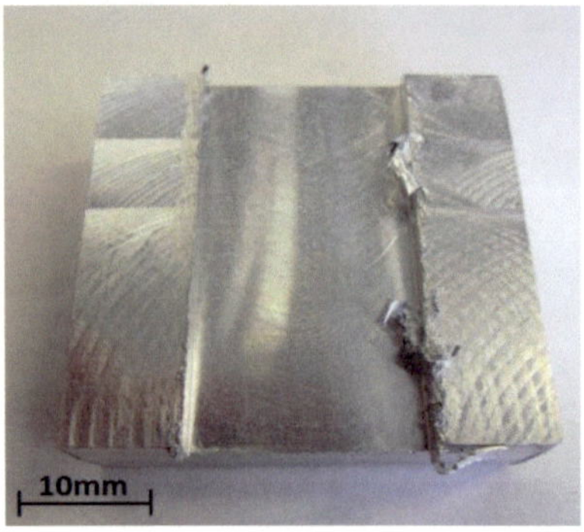

(a)

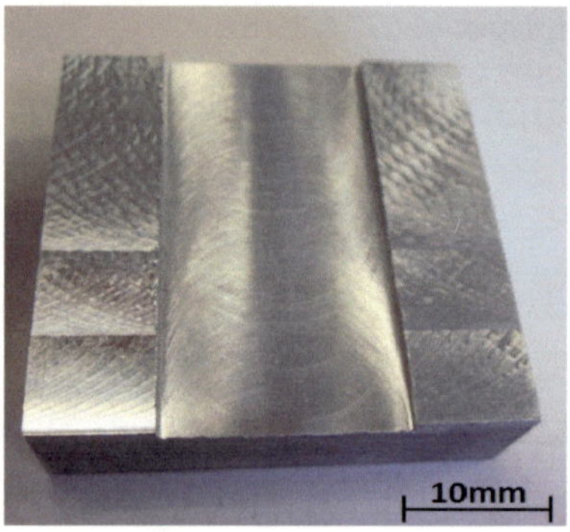

(b)

Figure 1. Slot-milled machined parts with (a) large burr formation, (b) burr formation with tiny scales

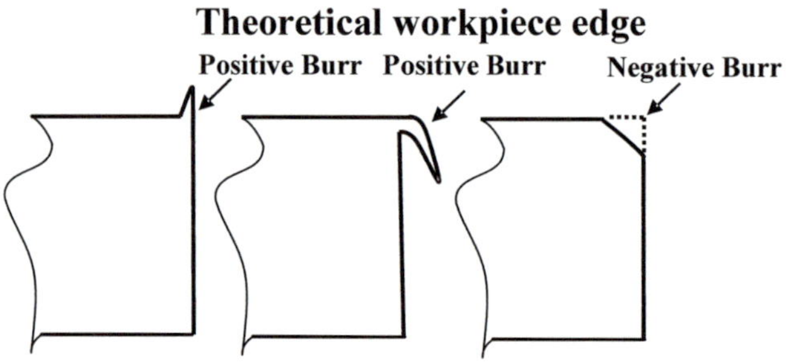

Figure 2. Examples of burr definition (*adapted from* [4])

Several authors have contributed to the advancement of knowledge on burr formation mechanisms, among which Gillespie [2], Aurich et al. [3], Pekelharing [5], Sofronas [6]Nakayama and Arai [7], Chern and Dornfeld [8], and Hashimura et al [9, 10]. Currently, numerous burr descriptions exist, depending on the application, manufacturing process, formation mechanism, shape and material properties [3]. The four main types of machining

http://dx.doi.org/10.5772/58361

burrs are the Poisson burr, the Rollover burr, the Tear burr and the Cut-off burr [4]. According to [11], Poisson burr is formed as a result of the material's tendency to bulge sidewise. Narayanaswami and Dornfleld [12] called this phenomenon a side burr, because, according to engineering mechanics, the *Poisson effect* is only present in the elastic range.

Two types of burrs known as primary and secondary burrs were introduced by Kishimoto et al.[13]. Beier [14] described a secondary burr as remaining material at the edge of a part after deburring process. From [3], secondary burrs formed after the breakage of the primary burrs. However, they are smaller than depth of cut, while primary burrs are larger [13]. Nakayama and Arai [7] described the burr formation in various machining processes by combining two classification systems as: [1] by direct concerning of cutting edge; [2] by mode and direction of burr formation. The various types of machining burrs are shown in Figure 3. Interested readers on different types of machining burrs are referred to [3].

(1) Cutting edge directly concerned	
Major cutting edge	M
Corner or minor cuttign edge	C
(2) Mode of the diection of formation	
Backward flow	B (Backward or entrance flow)
Sideward flow	S (Sideward burr)
Forward flow	F (Forward or exit burr)
Leaning to feed direction	L (Leaned burr)

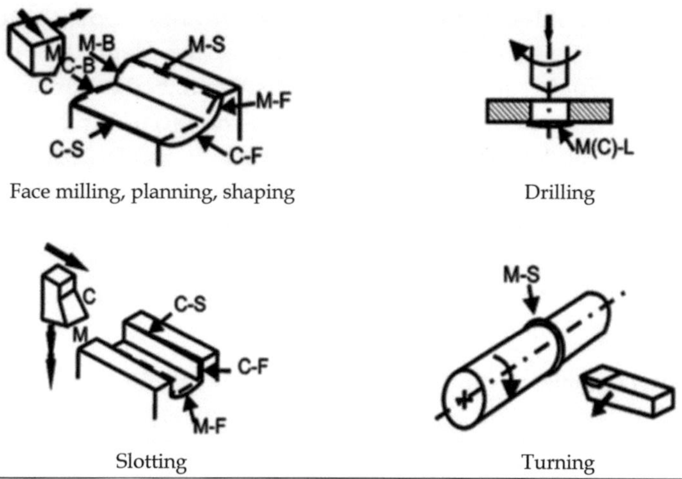

Figure 3. Types of machining burr [3]

As can be seen in Figure 4, to better describe the burr, a new term called "burr value" was defined in [15]. It contains the burr root thickness (b_r), burr height (b_h), burr thickness (b_t) and burr root radius (r_f). However, measuring and/or estimating all these parameters to calculate the burr value is very difficult and time-consuming. Furthermore, it would appear that the burr value can also not be used as an efficient parameter to better select a deburring method. Of all the burr parameters, the burr height and thickness are used to determine the burr removal difficulties [16].

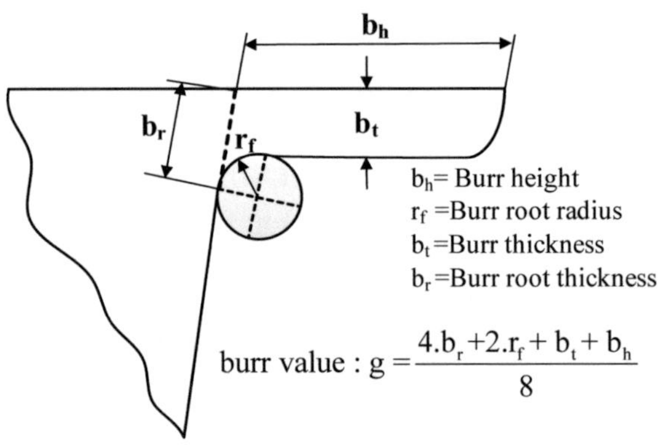

Figure 4. Measurement values of a burr (*adapted from* [15])

3. Factors governing burr formation

Burr formation is a crucial issue in industrial circles. Previous studies have shown that burr formation is almost impossible to avoid [17]. Gillespie and blotter [18] observed that burr formation cannot be avoided solely by changing the feed rate, the cutting speed and the tool geometry. Burrs in drilling perform an important role on product quality and may cause reliability problems and performance degradation. Burrs are formed both at the entrance and at the exit of the workpiece [19]. The exit burr is important as it is larger in size and is most difficult to remove causing deburring problems. Therefore, many of studies paid their attention to exit burrs in drilling and milling operations [19-27]. Sofronas [6] summarized several factors governing drilling burr formation. According to [3, 6, 28], the following are the principal factors governing milling burr formation:

1. Machined part (geometry, dimension, mechanical properties, etc.);
2. Cutting parameters (cutting speed, feed rate, depth of cut, etc.);
3. Cutting tool (material, shape, geometry, rake angle, lead angle, helix angle, etc.);
4. Machine tool (rotational speed, dynamic strength, etc.);
5. Manufacturing strategy (tool path, coolant, back cutting, lubrication, MQL, etc.).

This summary is however still limited due to the complex interaction effects that exist between process parameters, since their degree of influence on burr formation varies considerably simply by adding or removing cutting parameters and/or changing the material. In other words, factors governing burr formation cannot easily be classified as direct and indi-

http://dx.doi.org/10.5772/58361

rect factors [29]. The following lines present the dominant process parameters influencing the burr formation mechanism and size.

3.1. Workpiece materials and conditions

Machined part properties (e.g., chemical, mechanical properties) have significant effects on the burr formation process. The dominant mechanical properties usually reported in the literature are hardness, ductility, yield strength and elongation [30]. According to [3], higher ductility materials tend to generate larger burrs, but limited if not none burr formation is anticipated when the material is restricted to deform in the force direction. According to [31], the machining of ductile materials tends to form larger burrs, particularly at higher levels of cutting speed and feed rate. According to Ko and Lee [32], material properties had more effect on the drilling burr size than feed rate. Analytical models proposed in [27, 33] were capable to predict the size and type of exit burrs in milling and drilling of ductile materials. Niknam and Songmene [27], while modeling and studying the burr formation during millings of AA6061-T6 and AA2024-T321 found that the exit burr thickness, which control the deburring difficulties and the deburring cycle time, is highly sensitive to material mechanical properties such as yield strength and to the cutting force. When the material is brittle, fractured burrs (negative burrs) are formed on the edge part. This phenomenon can be reinforced at higher cutting speeds and feed rates, creating irregular burrs. The workpiece edge angle is the most prominent geometrical element of the workpiece that highly affects the burr formation mechanism. According to [34, 35], cutting tests on the edge angle lower than 90º generate long and thin burrs, while short and thick burrs are formed on parts with edge angles of 90º or larger. An increase in the temperature hardens most materials, and consequently affects the machining and deburring performance, even if the burrs created are small. According to [2], taking steps to prevent plastic deformation reduces the incidence of burr formation. His proposed methods include laser treatments, hard machining, localized mechanical processes, and chemical and thermal treatments. In addition, chamfering on the external edges of the machined part before the cutting operation is an excellent approach to prevent material deformation at the part edge, and consequently achieve burr size reduction [17]. The burr form and height in drilling are dependent on the material properties and cutting conditions [36]. Images showing typical exit hole appearance are presented in Figure 5 as a function of feed rate. As shown in the following example, the AA6061 High Strength (AA6061HS) contained higher levels of the major alloy addition to promote increased age hardening response. The AA6262 and AA4XXX compositions were based on AA6061 but contained additions of 0.5 wt% Pb-0.4 wt% Bi and 8% Si respectively.

The worst case for burr was obtained for the AA4XXX-T6 alloy (Figure 5] which is the most ductile one. The burr observed was a transient burr type. The other alloys exhibited a uniform burr (type I) or crown burr (type II):

- The AA6262-T6 and AA6061-T6HS alloys produced only uniform burrs (type II).
- The AA4XXX-T6 and AA6061-T6 alloys produced both uniform burrs (Type II) and transient or crown burrs (type I). The latter are generally difficult to remove.

- The AA4XXX-T6 was problematic in terms of exit burr height.

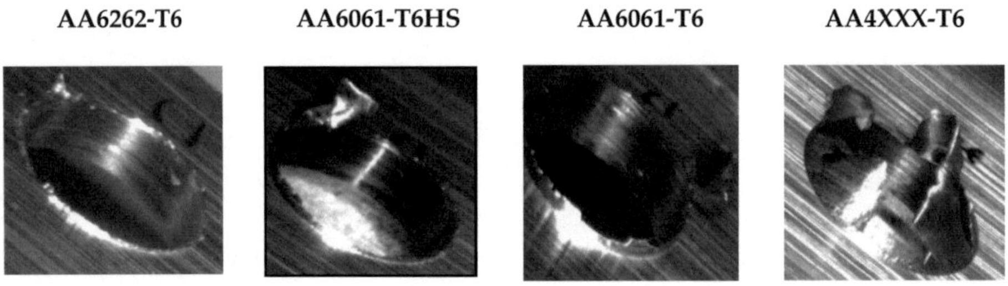

Figure 5. Optical microscopy images of exit burrs observed on drilled holes as a function of feed rate (cutting speed: 45.7 m/min; Feed rate: 0.0508 mm/rev) [36]

3.2. Cutting conditions

According to [37], burr height varies irregularly with changing cutting conditions. Increasing the cutting speed leads to reduced burr size. In addition, milling operations at higher feed rates reduces the burr size, while creating secondary burrs that are easier to remove. From [38], when the machined part surface is hardened in high speed machining, a transition from ductile to brittle behavior may occur. This phenomenon may lead to decreased burr height. Chern [39] analyzed burr formation during the face milling of aluminum alloys. He found that secondary burr formation is dominated by the depth of cut and the feed rate. Nakayama and Arai [7] showed that the burr size can be reduced by limiting the undeformed chip thickness. The cutting conditions, tool and workpiece geometry may reduce the shear strain supported by the chip, therefore possibly leading to burr reduction. Kim and Dornfeld [31] showed that higher levels of depth of cut generally increase the burr size. Longer burrs in the cutting direction are formed when larger corner radii are used. Olvera and Barrow [40] found that the exit angle and the depth of cut influence the exit burr in the cutting direction, whereas the depth of cut is the main factor affecting the exit burr in the feed direction. According to [41, 42], the use of high levels of axial depth of cut increases the possibility of burr size minimization, but may also cause inevitable damage to the cutting tool, the machine and machined part functionality. Therefore, the use of very high and/or low cutting parameters levels is not suggested during milling operations. Ko and Lee [32] used multiple materials in drilling processes and they concluded that the burr thickness is independent of the feed rate.

As shown in [36], the lower the feed rate, the higher the burr height obtained (Figure 6]. The AA4XXX produced most of the times high size burrs and only in very limited cases, the burr size was comparable to others alloys tested. At lower speeds, the burr size observed was higher compared the one obtained at high cutting speed; this denotes a possible interaction of the feed rate and the cutting speed on burr formation.

http://dx.doi.org/10.5772/58361

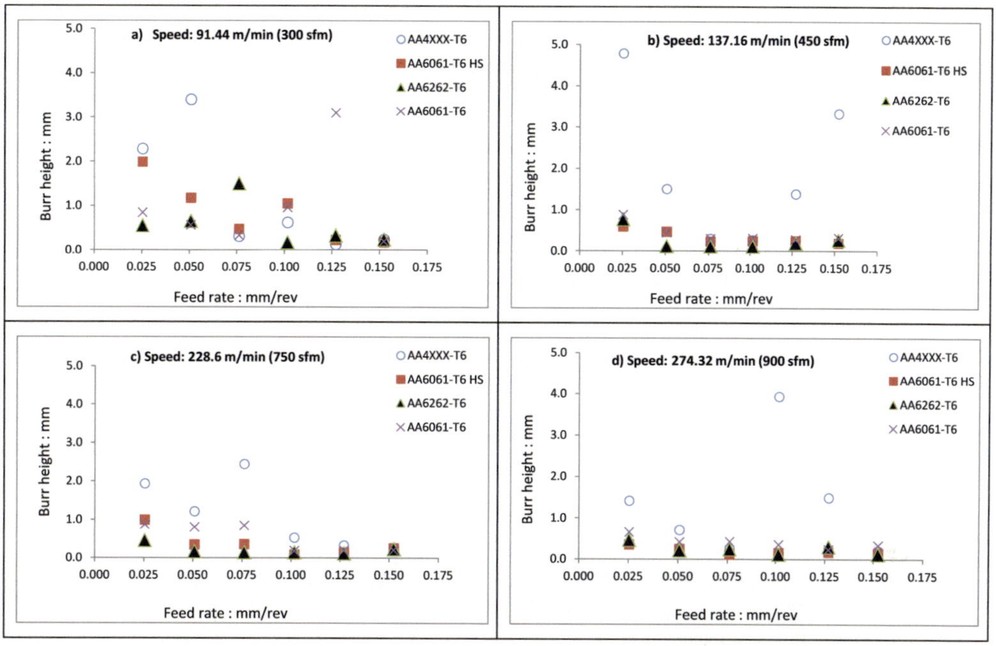

Figure 6. Burr height progression as a function of feed rate, cutting speed and workpiece materials [36]

The effects of cutting parameters on top, entrance and exit burr thickness and height during the slot milling of AA 2024-T351 and AA 6061-T6 were statistically investigated in [24, 25, 43]. Among the investigated burrs, exit up milling thickness could be controlled by cutting process parameters, such as feed per tooth, depth of cut and cutting tool (see Fig.7(a)). While other burrs, such as exit up milling burr height, are affected by interactive between process parameters, nor direct effects.

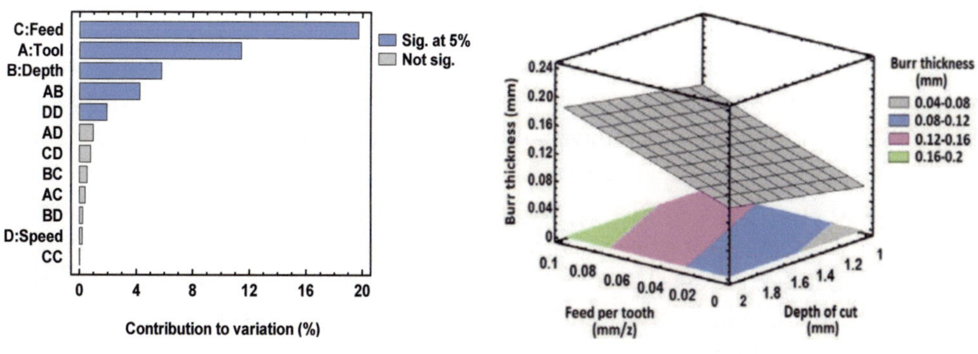

(a) (b)

Figure 7. (a) Pareto chart and 3D contour plot of exit up milling burr thickness of AA 6061-T6 [21]

It is believed that the burr formation mechanism depends highly on the chip formation mechanism [27]. Cutting forces and chip thickness $h(\varphi)$ are highly affected by the feed rate, the depth of cut, and the tool and workpiece geometry. Therefore, it could be inferred that burr formation is influenced by cutting forces [17]. The direction and intensity of cutting forces affect the volume of the chip generated, and can also play an important role in material deformation. The influence of cutting forces on drilling and milling burr formation has been reported in [27, 29, 44, 45]. According to [46], the variation of exit up milling burr thickness (B_t) is highly correlated with changes in tangential cutting force [Figure 8]. According [46], the burr thickness (B_t) and cutting force F_t could be linearly formulated as a function of depth of cut and feed per tooth.

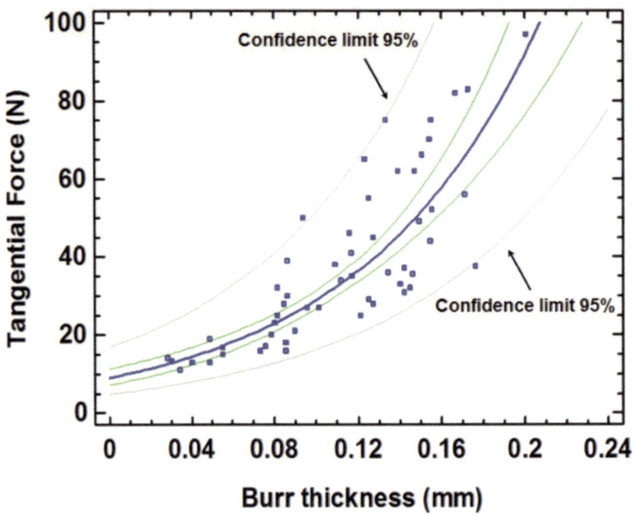

Figure 8. Exponential regression model between F_t and B_1 thickness [46]

Considering the research works presented in this section on factors governing burr formation, it could be inferred that due to complex mechanisms of burr formation and direct and interactive effects between process parameters, a large number of experiments is required to evaluate the effects of process parameters on burr formation and size [47]. The combination of statistical and experimental approaches is a good method to better understand the burr formation mechanism and to define the factors governing milling burrs. Furthermore, knowing that the best setting levels of process parameters needed to minimize each response are not similar, the question is how to obtain the best setting levels of process parameters to reach the optimum or near-optimum burr size. This issue becomes more complex as for a given machined part; cutting parameter optimization for burr size minimization alone may frequently deteriorate other machining performances, such as tool life and surface roughness. Therefore, the use of optimization methods for the correct selection of process parameters is strongly recommended [48]. According [21, 49, 50] achieving better surface finish and acceptable burr size in milling and drilling operations at the same time is possible when using optimization tools such as the desirability function and Taguchi method.

3.3. Cutting tool geometry

Pande and Relekar [51] studied the influence of drill diameter, feed, length to diameter ratio, and material hardness on burr height and thickness. It was determined that a drill tool diameter in range of 8-10 mm resulted in the low values of burr height. Bansal [52] found that using inserts with positive axial rake and negative radial rake angles led to satisfactory burr size and surface quality. According to [24], slot milling with a larger insert nose radius (Rε) leads to bigger exit bottom burr and smaller exit up milling side burr. In addition, when a larger Rε, pretty close to the axial depth of cut is used, a primary exit bottom burr formation is expected. Consequently, a smaller exit up milling side burr is generated. Avila and Dornfeld [37] showed that tool geometry and in-plane exit angle Ψ have significant effects on burr size and edge breakout during the face milling of aluminum-silicon alloys (AlSi9Cu3 and AlSi7Mg). Tripathi and Dornfeld [53] reported the possibility of burr-free conditions when using diamond end mill tools at high cutting speeds. According to [18], the use of sharp cutting edge tools with positive rake angle avoids built-up edge (BUE) formation, thus reducing the burr size. According to [42], tool coating has a negligible influence on face milling burrs. However, a certain level of coating influence on slot milling burrs was observed in [24]. According to [31, 54, 55], the tool condition and cutting parameters used, in particular, the feed rate, are the main governing factors affecting burr formation. For instance, a sharp cutting edge tool with a positive rake angle in the case of a milling operation avoids built-up edge formation, thus decreasing burr formation [18]. As presented in [54], tool wear increases the contact area of the burr-tool interface, and consequently, increases the cutting forces and stress distributions (Figure 9). Tool wear may physically occur on two sides of the cutting tool, mainly on the rake face and the flank face, thereby forming crater wear and flank wear. According to Choi et al. [56], tool wear highly affects the burr formation process when the tool enters and exits the machined part. Large entrance burrs formed when using worn tools resulted by different kinematic engagement rather than back cutting.

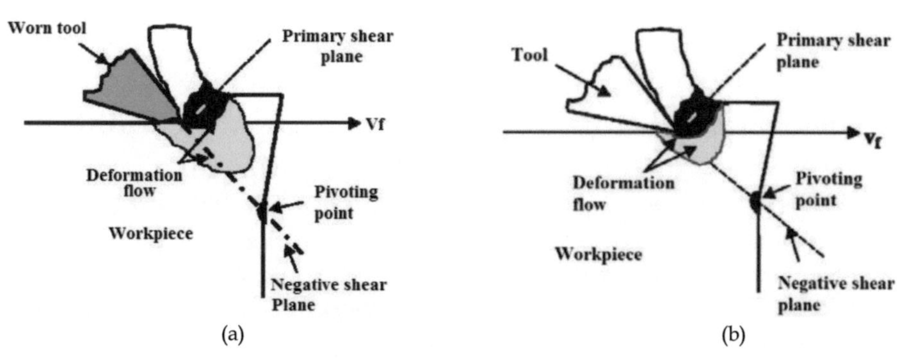

Figure 9. Shear angle and pivoting point for a (a) worn tool, (b) sharp tool (*adapted from* [54])

According to [18, 28, 31], using higher cutting speeds and feed rates when cutting certain materials may increase the cutting temperature and consequently reduce the tool life as a result of tool wear. When the strain rate of the material increases, it effectively enlarges the burr dimension. A similar problem may occur when using dry machining or inadequate tool geometry.

3.4. Machining strategy

According to [55, 57], an adequate selection of the machining strategy has positive effects on the burr formation mechanism. The main machining strategies proposed to date include:

1. Optimization of the tool path planning, including the machining direction and the tool engagement angle;
2. Using inserts and backup materials;
3. Using modified cutting parameters;
4. Using coolant and lubrication.

To predict and control the burr size in milling, several algorithms were presented in [26, 58, 59] and eventually led to proposing tool path planning approaches for burr size minimization. The effect of back cutting on burr formation was reported in [60]. Przyklenk [34] proposed a new strategy for burr reduction by using dry ice snow to cool down the machined part edge. Shefelbine and Dornfeld [41, 61] stated that the use of coolant decreases the burr size, while larger burr is expected when using worn tools. According to [17], proper lubrication reduces friction between the workpiece and tool, and consequently, reduces the incidence of burr formation. However, as stated by Aurich et al.[62], in the case of certain materials, the use of lubricant hardens the burrs and complicates the deburring processes. Moreover, the use of cutting fluids seriously degrades the environmental air quality and increases machining costs by 16-20% [36]. One alternative approach with reduced cost and greater environmental benefits is dry machining. Some works have reported on dry milling [21, 43, 63-67] and dry, mist and lubricated drilling of aluminium alloys [29, 49, 68]. According to [69], the processes constitutes a suitable candidate for the machining of ductile materials, such as aluminum alloys.

4. Case study: Effect of artificial aging heat treatment on drilling burr size

Surprisingly, very limited information is available on the influence of material properties such as ductility, strength, and hardness on the burr formation during machining of aged aluminum alloys. In this chapter, the influence of heat treatment on the burr formation during drilling of Al-Si-Mg (A356] cast alloys is investigated. In order to study this effect systematically, the aluminium alloys were heat-treated to produce different precipitation states they were later machined under controlled conditions. This work was carried out in order improve part quality of drilled hole.

4.1. Experimental procedure

A set of experiments were carried out on a high speed 3-axis CNC milling machine-tool (Power: 50 kW, Speed: 28000 rpm, Torque: 50 Nm). The cutting tool used is non-coated high speed steel twist drills (3/8 stub drill bright finish with 118 point angle). It should be mentioned here that a group of drills with the same batch were used throughout the tests in order to ensure uniformity of geometry, microstructure and properties for the cutting tools. The parts used for the experiment were rectangular blocks of A356 cast aluminum alloy 300 x100 x 20 mm in size, mounted on a special machining fixture. Cutting forces were measured using a three-axis table dynamometer (Kistler 9255-B). The tested material was A356 aluminum alloy which chemical composition is given in Table 1. Drilling tests were conducted at different cutting conditions as shown in Table 2. Experiments were repeated four times and the average values of burr sizes were used for further analysis. The burr height was recorded using Mitutoyo Height Gauges with a sensitivity of 0.0005 in (13 μm).

	Si %	Mg %	Fe %	Cu %	Mn %	Zn %	Al
A356	7	0.35	0.2	0.2	0.1	0.1	Balance

Table 1. Chemical composition of A356 alloy

Parameters	**Condition**
Material	A356 (300 mm x100 mm x 20 mm)
Tool	HSS twist drill-9.525 mm diameter , 118° point angle
Speed	60, 180, 300 m/min
Feed	0.15, mm/rev
Depth of cut	3 mm
Lube	None

Table 2. Machining parameters used

5. Results

5.1. Effect of heat treatment

The properties of aluminum casting alloys can be improved through the appropriate control of several metallurgical factors involved in the production of these castings [70]. Aging treatment is usually applied to improve the strength and hardness of the castings. Aging temperature and aging time are the two variables which control the characteristics of the phases precipitated during the aging treatment, and they ultimately also control the mechanical and machinability properties of the alloys. The A356 alloys were received in as-cast condition (T0), and the samples were then divided into five groups as follows: (i) two blocks in as-cast condition (T0); (ii) two blocks in T4 condition (Solution Heat-Treated " SHT "+Quenching); (iii) two blocks in T61 condition (SHT+Quenching+Artificial aging at 155°C for 5 hours); (iv) two block in T62 condition (SHT+Quenching+Artificial aging at 180°C for 5 hours); and (v) two blocks in T7 condition (SHT+Quenching+Artificial aging at 220°C for 5 hours). All the samples were solution heat-treated at 540˚C, for solution times of 8 hours. The solution-treated samples were then quenched in warm water (60˚C) to room temperature. For each given condition, all the samples were solution heat-treated and quenched at the same time leaving only the other conditions, such as natural and artificial aging time, as variables as shown in Figure 10. Figure 11 summarizes the hardness measurement implemented on the A356 aluminum alloy. The hardness measurements reveal that the peak hardness value varies between 85HRE and 90 HRE. Hence, it is believed that aging at 155˚C for 5 hours produces almost the same precipitation hardening as aging at 180˚C for 5 hours. Then hardness tends to decrease upon further aging. The variations of hardness when exposed to different aging temperatures are correlated with the number of Mg_2Si phases in which the hardness increases with an increase in the number of Mg_2Si phases.

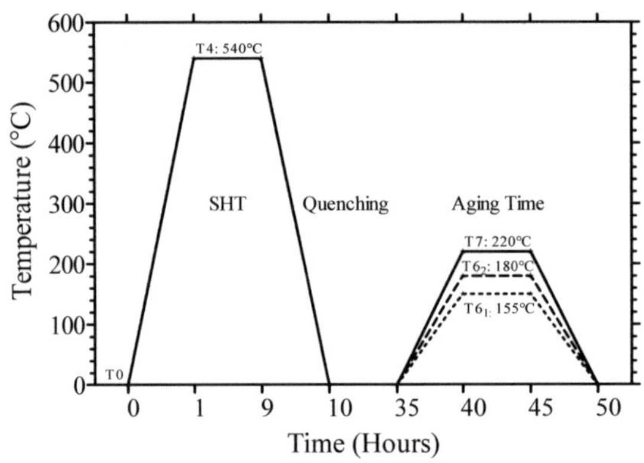

Figure 10. Different heat treatment conditions applied to the A356 alloy

Figures 12-13 show the effects of different heat treatments on the burr heights with various cutting speed and feed rate during machining A356 aluminum alloys in dry conditions. To emphasize the effect of heat treatments on the material properties and its effects on the burr height thus, the case study will be focused on dry conditions. The experimental results reveal that the A356-T61 aluminum alloy in peak aging condition produces the lowest burr height than other alloys, whereas the A356-T0 and A356-T4 in cast and SHT conditions were produced the highest level of burr formation. Generally, it is also observed that aging at low temperature, 155 ºC, was observed to produce the lowest level of the burr formation while the aging at higher temperatures, 180°C, and 220°C, respectively, are accompanied with an increase in the burr heights. The lowest level of burr height was observed in peak aging conditions A356-T61 alloy may be explained based on the fact that burr formation is closely re-

http://dx.doi.org/10.5772/58361

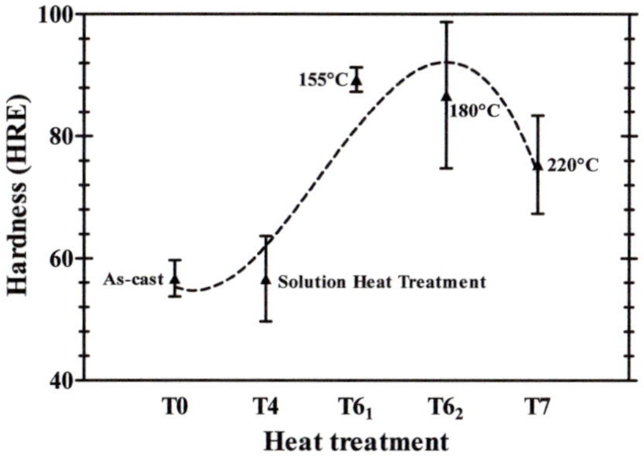

Figure 11. Evolution of the changes in the hardness values result in the heat treatment conditions

lated to the hardness of the workpiece, as well as to the deformability of the matrix around the Si particles. Thus, when the workpiece made of the A356-T61 alloy is hard (i.e. high tensile properties), thereby resisting deformation, and the work material is also difficult to deform in the shear zone ahead of the cutting edge and on the tool surface in such a way that the Si particles are secured firmly in position. Consequently, the particles have a strongly abrasive action on the tool cutting edge resulting in a high cutting force as well as in low level of burr formation.

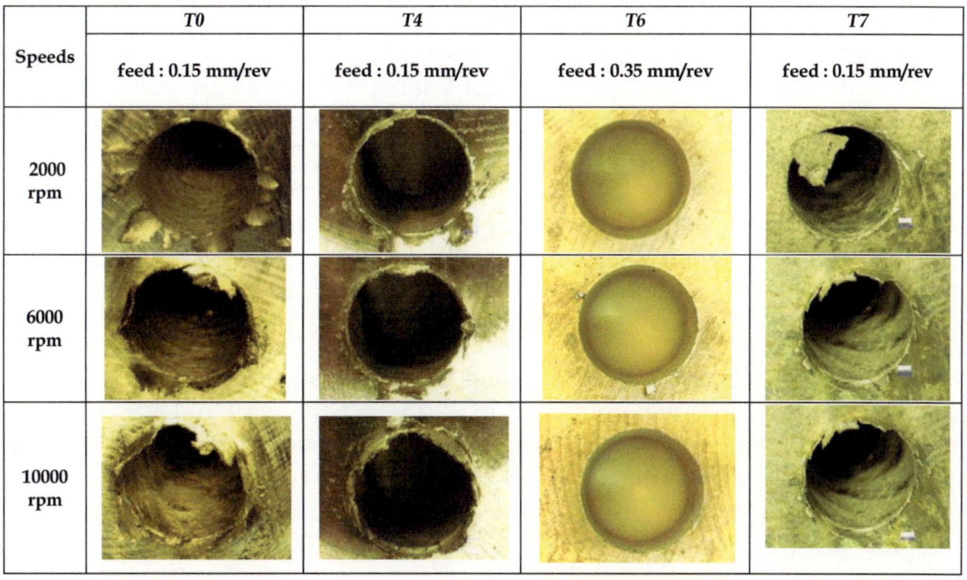

Figure 12. Influence of the heat treatments on the burr formation during the drilling of the A356 aluminum alloy with diffiren feed rate at (a) as-cast alloy condtion (T0), (b) T4 (c) T61 (d) T62 (e) T7 conditions

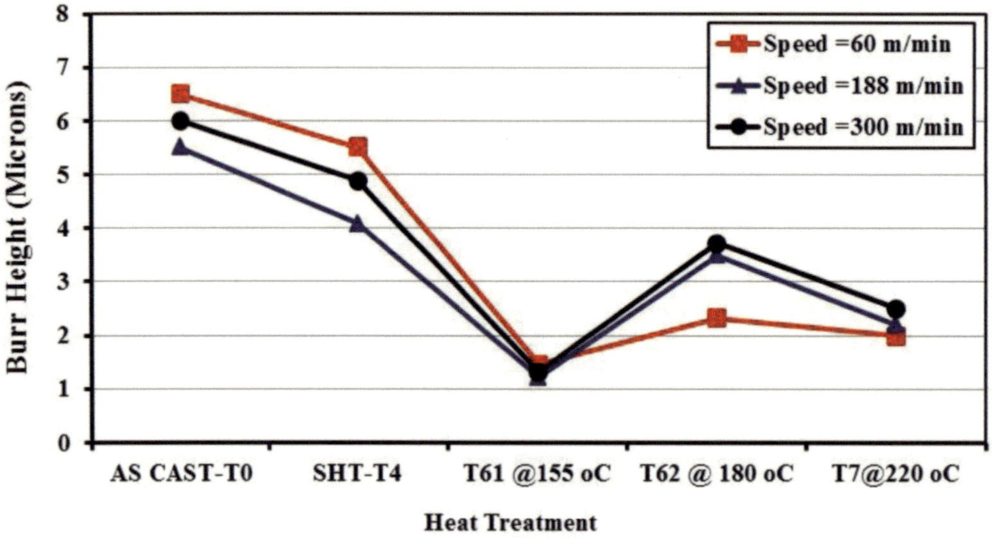

Figure 13. Influence of the heat treatments (T0, T4, T61, T62 and T7) and the cutting speed on the burr height during the dry drilling of the A356 alloy with feed rate 0.15 mm/rev.

6. Overview of deburring processes applicable to aluminium alloys

Burr removal is a non-value added process [3] and might represent as much as 30 percent of the cost of finished parts [2]. As deburring is non-productive and costly finishing process, it should be minimized or avoided. Any material leading to limited burr formation is therefore advantageous. Recent studies and literature have pointed out tremendous issues related to burr formation and deburring operations, including: [1] small finger injuries for assembly workers; [2] source of debris (bits of burrs) during operation, thereby reducing the life time of the machined part; [3] changing parts resistance and reduction of tool life and efficiency [44]; [4] presentation of hazard in handling of machined parts, which can interface with subsequent assembly operations ; and [5] the burrs that are adhered to the work part may become loose during operation, and consequently cause difficulties and damage.

Gillespie [2] has identified 122 deburring and edge finishing processes. To better select them, several classifications were proposed in [2, 15, 34]. The work of Gillespie [2] is the most complete work, encompassing all deburring methods, from manual deburring to high technology finishing systems using CNC and industrial robots. He [2] classified deburring processes under following categories of [1] Mechanical deburring processes; [2] Thermal deburring processes; [3] Chemical deburring processes and [4] Electrical deburring processes.

Mechanical deburring processes: During mechanical deburring processes, the burrs are reduced or removed by mechanical abrasion. This can be done manually, using abrasive, using a brush or a solid tool off-line or directly at the machine-tool station. Sometimes, a robot

http://dx.doi.org/10.5772/58361

is also used. The overview of most highly used mechanical deburring methods on aluminum alloys will be presented in the following sections:

6.1. Manual deburring

Manual deburring is still known as the most widely used operation for many reasons, including extreme flexibility, low cost and lack of technology needed. According to [2], manual deburring is associated with wasting of time and asset, fatigue, frustration, etc. Moreover, in most of industrial sectors, manual deburring is implemented in dry conditions by non-qualified operators. This consequently increases the waste rate and delay in production lines.

6.2. Bonded-abrasive deburring

Bonded abrasive deburring *or sanding* is a versatile deburring technique which can be applied when heavy stock removal is intended. This method performs well in manual and automated operations that are used for deburring and surface smoothing. Many types of bonded abrasives are available for dry and lubricated deburring of aluminum alloys and metals work parts. The main benefits of bonded-abrasive deburring are low cost; large variety of models and great adaptability to manual or automatic equipment. On the other hand, the main disadvantages of this method include short life time, dust emission and new burrs, significant effects on residual stress and surface quality and lack of access to certain sides of the work part.

6.3. Brush deburring

The power driven brush tools have a wide range of applications in deburring, cleaning, descaling, polishing, edge blending and texturizing of the metal work parts. The brush deburring is considered as a fast, safe, simple, relatively inexpensive, and flexible deburring method, which could be also adaptable to manual or automatic equipment with little operator interference. The rotary action in brushing allows a great variety of driving motor and fixtures to be employed. The brush deburring involves several environmental, health and safety considerations, including particle and dust emission when using dry sanding of metal and plastic parts. The generation of new burr, new changes on the work part size, fatigue life and residual stress are the main disadvantages and side effects of brush deburring. As described in [2], brush deburring method is widely used for deburring of aluminum work parts, such as cylinder heads (see Figure 14). The main process variables involved in brush deburring include, brush style, brush design and materials, face width, coolant, brush rotational speed, burr size, burr location and work part material.

6.4. NC/CNC machining centers

Due to growing demands on higher production rate, improved quality and less labour and production cost, particular interest has been paid to use of NC/CNC machines for precise deburring and chamfering of holes, flat and curved surfaces. The NC/CNC machines can

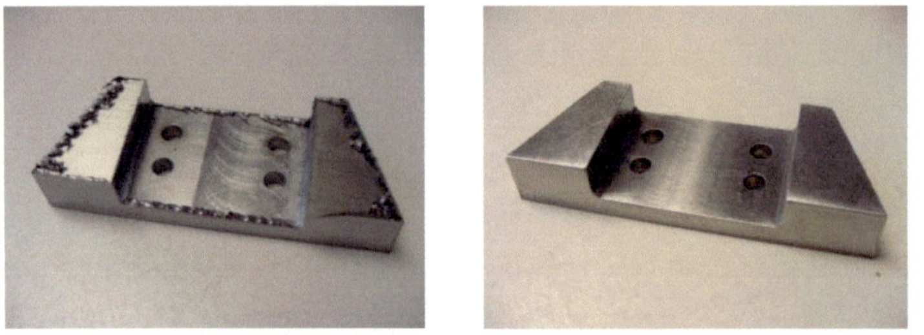

(a) Before deburring (b) After deburring

Figure 14. Burr edges before (a) and after brush deburring(b) of AA 6061-T6 [71]

brush the machined parts by simply attaching the brushing tools (miniature or large scales) in a tool holder. It also allows the machine to change the tooling conditions and begins the cutting operations and simultaneously taking the advantage of over 1000 standard cutting tools, thus providing great flexibility. Other benefits of NC/CNC machines include prevention of repetitive motions in hand deburring, and lost time due to work-related injuries which may lead to a major cost saving in production line [2]. When using NC/CNC machines, it is also possible to pick up a movable water jet nozzle and traverse it around the machined part edges for deburring and edge finishing (see Figure 15). However this method can be used on the aluminum work parts which require reasonable but not complete burr removal [3]. A polishing/deburring machine is developed in [72], consisting of two subsystems. The first subsystem is a five-axis machine for tool/part motion control and the second subsystem is a compliant tool head for tool force control. Both subsystems are designed based on the tripod principle. According to experimental results, high precision automated polishing/deburring on aluminium work parts was observed. As pointed out in [3], NC/CNC machines may not produce high quality cast or forged surfaces. The main concerns when using NC/CNC machines are comprehensively presented in [2].

6.5. Robotic deburring

Robots can operate with no time limit; reproduce the same motions accurately; can process workpieces faster than humans; they can use heavier; higher-powered tools for faster finishing; they can work in hazardous; noisy and ergonomically unsuitable situations for humans. Robotic deburring is used to reduce the work load and guarantee an adequate workpiece quality level. Robotic applications fall into three general areas [1] simple-shape deburring and chamfering, [2] contouring and [3] sensor-controlled countering. A framework for robotic deburring applications in various industrial sectors was proposed in [73]. The use of robots for deburring operation was reported in [74, 75]. Robotic deburring of gearbox casting made from aluminum alloys is presented in [76]. In [77], an on-line industrial robot path generation method has been developed and implemented to generate robot paths for deburring cast aluminum wheels. This method could automatically generate six degree of freedom (DOF) tool paths for an accurate and efficient deburring process. Kazerooni [78] presented robotic deburring using tungsten cemented carbide rotary files. He introduced ro-

http://dx.doi.org/10.5772/58361

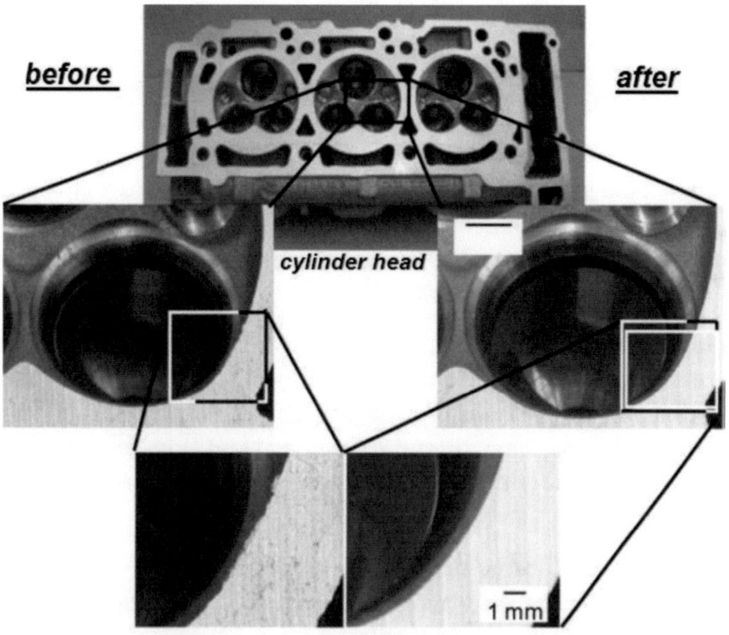

Figure 15. High pressure water jet deburring of aluminum cylinder heads [3]

bot-position uncertainties in deburring and a feed-back system working according to the prescribed controlled strategy. Experimental verifications on aluminum work parts have shown satisfactory results. Dornfeld [79] introduced the fundamental principles of acoustic emission (AE) applications in chamfering and deburring operations and verified his approach through experimental works on 6061-T6 aluminum alloys. Hirabayashi *et al.* [80] presented deburring robots equipped with force sensors for automatic deburring of elevator guide rails. The more widely used applications employ advanced robots that use five-axis compliant tools, capable to remove most, but not all burrs [2].

7. Conclusion

The development of aluminum alloys is often conditioned by aeronautical requirements, but still aluminum is considered as a suitable candidate for several applications in other sectors. However, through growing demands on part quality, functional performance and global competition, special attention has been paid to burr formation which appears to be one of the major troublesome impediments to machinability, part quality and high productivity of aluminium alloys.

In this chapter, overview of machining burr classifications and formation mechanisms as well as factors governing burr formation were presented. Furthermore, the effects of heat treatments on burr formation during drilling of aluminium alloys were presented, followed by introducing the most highly used mechanical deburring processes on aluminum work parts.

This work has led us to conclude that:

- There is a substantial need to reduce, prevent and eliminate burrs. Considering that the burr formation is inevitable phenomenon in machining operations, particular attention should be paid to burr control rather than burr avoidance. The main benefit would be to reduce the needs of deburring operations. However, using deburring operations in some cases is still mandatory.
- Most aluminium alloys, whether wroughts or casts, can experience burr formation during machining processes; The shape and the size of this burr will depend on the alloy composition and conditions, its mechanical properties, but also on type of machining operation, tooling used, machining parameters, and machining conditions and strategies. Using very low feed rates on a material with high ductility may generally lead to higher burr size.
- Development of simulation models of burr formation processes, coupled with advanced cutting force and temperature modeling algorithms, capable of indicating the interaction and dependencies of factors governing burr formation is suitable approach for better understating of factors governing each individual machining operation.
- The knowledge of each deburring method and the requirements of the finished products, in addition to burr size, mainly burr thickness are major parameters for correct selection of deburring method. For many cases, combination of several deburring processes is required to gain better results. Therefore, developing the links between burr sizes and deburring methods and deburring difficulty is a benefial approach for better selection of deburring methods, which infact reduce the non-desirable expenses.

Acknowledgements

This work has received financial supports from the Natural Sciences and Engineering Research Council of Canada (NSERC), consortium for Research and Innovation in Aerospace in Quebec (CRIAQ), and Fonds Québécois de la Recherche sur la Nature et les Technologies (FQRNT) by the intermediary of the Aluminium Research Centre Canada-REGAL.

Author details

Seyed Ali Niknam, Yasser Zedan and Victor Songmene*

*Address all correspondence to: victor.songmene@etsmtl.ca

École de Technologie Supérieure (ÉTS), Department of Mechanical Engineering, Montreal Quebec, Canada

References

[1] Karnik S, Gaitonde V, Davim J. A comparative study of the ANN and RSM modeling approaches for predicting burr size in drilling. The International Journal of Advanced Manufacturing Technology. 2008;38(9):868-83.

[2] Gillespie L. Deburring and edge finishing handbook. Dearborn, MI,USA: Society of Manufacturing Engineers (SME); 1999.

[3] Aurich JC, Dornfeld D, Arrazola PJ, Franke V, Leitz L, Min S. Burrs-Analysis, control and removal. CIRP Annals-Manufacturing Technology. 2009;58(2):519-42.

[4] Gillespie L. The battle of the burr: new strategies and new tricks. Manufacturing Engineering(USA). 1996;116(2):69-70.

[5] Pekelharing A. The exit failure in interrupted cutting. Annals of the CIRP. 1978;27(1): 5-10.

[6] Sofronas AS. The formation and control of drilling burrs. PhD Thesis, University of Detroit, USA; 1975.

[7] Nakayama K, Arai M. Burr formation in metal cutting. CIRP Annals-Manufacturing Technology. 1987;36(1):33-6.

[8] Chern GL, Dornfeld DA. Burr/breakout model development and experimental verification. Journal of Engineering materials and technology. 1996;118:201-6.

[9] Hashimura M, Hassamontr J, Dornfeld D. Effect of in-plane exit angle and rake angles on burr height and thickness in face milling operation. Journal of Manufacturing Science and Engineering. 1999;121(1):13-9.

[10] Hashimura M, Dornfeld D. Analysis of burr formation mechanism in machining process. Technical Paper, Society of Manufacturing Engineering (SME)-All series. 1999;121(1):1-7.

[11] Nisbet TS, Mullet G. Rolling Bearings in Service: Interpretation of Types of Damage: Hutchinson; 1978.

[12] Narayanaswami R, Dornfeld D. Design and process planning strategies for burr minimization and deburring. Transactions of the North American Manufacturing Research Institute of SME. 1994; 22:313-22.

[13] Kishimoto W, Miyake T, Yamamoto A, Yamanaka K, Takano K. Study of Burr Formation in Face Milling. Conditions for the Secondary Burr Formation. Bull Jpn Soc Precis Eng. 1981;15(1):51-2.

[14] Beier HM. Handbuch Entgrattechnik: Wegweiser zur Gratminimierung und Gratbeseitigung für Konstruktion und Fertigung: Hanser; 1999.

[15] Schäfer F. Entgraten: Krausskopf; 1975.

[16] Rangarjan A. Optimization of face milling process-Tool path and process planning techniques: Ph.D Thesis, University of California at Brekeley, USA; 2005.

[17] Tiabi A. Formation des bavures d'usinage et finition de pieces M.Sc Thesis, École de technologie superieure, Montreal, Canada; 2010.

[18] Gillespie L, Blotter P. Formation and properties of machining burrs. J Eng Ind(Trans ASME, B). 1976;98(1):66-74.

[19] Saunders LK, Mauch CA. An exit burr model for drilling of metals. Journal of manufacturing science and engineering. 2001;123(4):562-6.

[20] Lauderbaugh L. Analysis of the effects of process parameters on exit burrs in drilling using a combined simulation and experimental approach. Journal of Materials Processing Technology. 2009;209(4):1909-19.

[21] Niknam SA, Kamguem R, Songmene V. Analysis and optimization of exit burr size and surface roughness in milling using desireability function ASME 2012 International Mechanical Engineering Congress & Exposition IMECE 2012; 9-15 November 2012; Huston, TX, USA.

[22] Hassamontr J, Blondaz L, Dornfeld D, editors. Avoiding exit burr in CNC end milling by an adapted tool path. Symposium on Concurrent Design of Product and Manufacturing Processes ASME; 1998.

[23] Chu C-H, Dornfeld D. Tool path planning for avoiding exit burrs. Journal of Manufacturing Processes. 2000;2(2):116-23.

[24] Niknam SA, Songmene V. Factors governing burr formation during high-speed slot milling of wrought aluminium alloys. Proceeding of the Institution of Mechanical Engineers, Part B: Journal of Engineering Manufacture. 2013;227(8):1165-79.

[25] Niknam SA, Songmene V, editors. Experimental Investigation and Modeling of Milling Burrs. ASME 2013 International Manufacturing Science and Engineering Conference collocated with the 41st North American Manufacturing Research Conference; 6-10 June 2013; Madison, WI, USA.

[26] Niknam SA, Wygowski W, Balazinski M, Songmene V. Milling Burr Formation and Aavoidance. In: Davim JP, editor. Machinability of Advanced Materials. London, UK: ISTE Wiley; 2014. p. 57-94.

[27] Niknam SA, Songmene V. Modeling of burr thickness in milling of ductile materials. The International Journal of Advanced Manufacturing Technology. 2013;66(9): 2029-39.

[28] Tseng PC, Chiou IC. The burrs formation prediction and minimization based on the optimal cutting parameters design method. JSME International Journal Series C. 2003;46(2):779-87.

[29] Zedan Y, Songmene V, Khettabi R, Kouam J, Masounave J, editors. Surface integrity of Al6061-T6 drilled in wet, semi-wet and dry conditions. In Proceedings of the 37th

International MATADOR Conference, Manchester, UK, July 25-27, 2012:131-134. Springer Science and Business Media, LLC.

[30] Leopold J, Wohlgemuth R. Modeling and Simulation of Burr Formation: State-of-the-Art and Future Trends. Burrs-Analysis, Control and Removal. 2010:79-86.

[31] Kim J, Dornfeld DA. Development of an analytical model for drilling burr formation in ductile materials. Journal of Engineering materials and technology. 2002;124(2): 192-8.

[32] Ko S-L, Lee J-K. Analysis of burr formation in drilling with a new-concept drill. Journal of materials processing technology. 2001;113(1):392-8.

[33] Segonds S, Masounave J, Songmene V, Bès C. A simple analytical model for burr type prediction in drilling of ductile materials. Journal of Materials Processing Technology. 2012; 213(6): 971–977

[34] Przyklenk K. Abrasive flow machining–a process for surface finishing and deburring of work pieces with a complicated shape by means of abrasive laden media. Advances in Non-traditional Machining, ASME, PED. 1986;22:101-10.

[35] Wygowski W. Strategie de limitation de bavures pendant l'usinage. Ph.D. Thesis, Ecole Polytechnique, Montreal, Canada; 2004.

[36] Songmene V, Kouam J, Zaghbani I, Parson N, Maltais A. Global Machinability of Al-Mg-Si Extrusions. In: Ahmad Zaki P, editor. Aluminium Alloys-New Trends in Fabrication and Applications: InTech; 2013.

[37] Avila MC, Dornfeld DA, editors. On The Face Milling Burr Formation Mechanisms and Minimization Strategies at High Tool Engagement. Intl Conf on Deburring and Edge Finishing; 2004; University of California at Berekley.

[38] Rangarajan A. Priority Based Tool Path Planning for Face Milling. MSc Thesis, University of California at Berkeley, USA; 2001

[39] Chern GL. Analysis of Burr Formation and Breakout in Metal Cutting: PhD Thesis, University of California at Berkeley, USA; 1993.

[40] Olvera O, Barrow G. Influence of exit angle and tool nose geometry on burr formation in face milling operations. Proceedings of the Institution of Mechanical Engineers, Part B: Journal of Engineering Manufacture. 1998;212(1):59-72.

[41] Shefelbine W, Dornfeld D, editors. Influences on Burr Size During Face-Milling of Aluminum Alloys and Cast Iron. Consortium on Deburring and Edge Finishing, Laboratory for Manufacturing and Sustainability; 2004: University of California at Berkeley, USA.

[42] Olvera O, Barrow G. An experimental study of burr formation in square shoulder face milling. International Journal of Machine Tools and Manufacture. 1996;36(9): 1005-20.

[43] Niknam SA, Songmene V. Statistical investigation on burrs thickness during milling of 6061-T6 aluminium alloy. CIRP 1st International Conference on Virtual Machining Process Technology; 28 May-1 June 2012; Montreal, QC, Canada.

[44] Ko S, Dornfeld D. A study on burr formation mechanism. Journal of Engineering materials and technology. 1991;113:75-87.

[45] Niknam SA, Tiabi A, Kamguem R, Zaghbani I, Songmene V. Milling Burr Size Estimation Using Acoustic Emission and Cutting Forces. Proceedings of the ASME 2011 International Mechanical Engineering Congress & Exposition IMECE2011; 11-17 November 2011; Denver, Col, USA.

[46] Niknam S. Burrs Understanding, Modeling and Optimization during Slot Milling of Aluminium Alloys Ph.D. Thesis, École de Technologie Superieure, Universite du Quebec , Montreal, Canada; 2013.

[47] Niknam SA, Songmene V. Milling Burr Formation, Modeling and Control: A Review institution of Mechanical Engineers, Part B: Journal of Engineering Manufacture. 2014 (In press).

[48] Niknam SA, Songmene V. Simultaneous optimization of burrs size and surface finish when milling 6061-T6 aluminium alloy. International Journal of Precision Engineering and Manufacturing. 2013;14(8):1311-20.

[49] Zedan Y, Niknam SA, Djebara A, Songmene V. Burr Size Minimization When Drilling 6061-T6 Aluminum Alloy. ASME 2012 International Mechanical Engineering Congress and Exposition; 9-15 November 2012; Houston, TX, USA.

[50] Niknam SA, Songmene V. Simultaneous optimization of burrs size and surface finish when milling 6061-T6 aluminium alloy. International Journal of Precision Engineering and Manufacturing 2013;14(8):1311-20.

[51] Pande S, Relekar H. Investigations on reducing burr formation in drilling. International Journal of Machine Tool Design and Research. 1986;26(3):339-48.

[52] Bansal A. Burr prediction system for face milling operation. : Masters Thesis, University of California at Berkeley, USA; 2002.

[53] Tripathi S, Dornfeld DA. Review of geometric solutions for milling burr prediction and minimization. Proceedings of the Institution of Mechanical Engineers, Part B: Journal of Engineering Manufacture. 2006;220(4):459-66.

[54] AM De Souza J, Sales W, Ezugwu E, Bonney J, Machado A. Burr formation in face milling of cast iron with different milling cutter systems. Proceedings of the Institution of Mechanical Engineers, Part B: Journal of Engineering Manufacture. 2003;217(11):1589-96.

[55] Wang GC, Zhang CY. Mechanism of Burr Formation in Milling. Key Engineering Materials. 2003;259:278-81.

[56] Choi GS, Wang ZX, Dornfeld D, Tsujino K, editors. Development of an intelligent on-line tool wear monitoring system for turning operations. Proceeding of Japan-USA symposium on Flexible Automation , ISCIE, 53, Kyoto, Japan, 1990.

[57] Chu C, Dornfeld D. Linking tool paths generated with different offset distances for edge quality enhancement in planar milling. Proceedings of the Institution of Mechanical Engineers, Part B: Journal of Engineering Manufacture. 2004;218(7):721-30.

[58] Chu CH, Dornfeld D, Brennum C. Prediction and simulation of milling burr formation for edge-precision process planning. University of California at Berekley, USA, 2000.

[59] Rangarajan A, Chu CH, Dornfeld D. Avoiding tool exit in planar milling by adjusting width of cut. ASME MFG Eng Div. 2000;11:1017–27.

[60] Rangarajan A, Dornfeld DA. Back Cutting and Tool Wear Influence on Burrs in Face Milling-Analysis and Solutions. Consortium on Deburring and Edge Finishing, Laboratory for Manufacturing and Sustainability,UC Berkeley. 2004:63-70.

[61] Shefelbine W, Dornfeld DA, editors. The Effect of Dry Machining on Burr Size. Consortium on Deburring and Edge Finishing, Laboratory for Manufacturing and Sustainability, University of California at Berkeley, USA; 2004.

[62] Aurich J, Sudermann H, Bil H. Characterisation of burr formation in grinding and prospects for modelling. CIRP Annals-Manufacturing Technology. 2005;54(1):313-6.

[63] Niknam SA, Zedan Y, Songmene V. Burr formation during milling of wrought aluminum alloys. 20th ISME Annual International Conference on Mechanical Engineering; 16-18 May 2012; Shiraz, Iran.

[64] Gillespie LRK. Deburring technology for improved manufacturing. Dearborn, MI, USA: Society of Manufacturing Engineers (SME); 1981.

[65] Kim JD, Kang YH. High-spend machining of aluminium using diamond endmills. International Journal of Machine Tools and Manufacture. 1997;37(8):1155-65.

[66] Dasch JM, Ang CC, Wong CA, Cheng YT, Weiner AM, Lev LC, et al. A comparison of five categories of carbon-based tool coatings for dry drilling of aluminum. Surface and Coatings Technology. 2006;200(9):2970-7.

[67] Songmene V., Khettabi R, Kouam J. High Speed Machining: A Cost Effective & Green Process. Int J Manufacturing Research (IJMR). 2012;7(3):229-56.

[68] Zedan Y, Songmene V, Kouam J, Masounave J. Effects of lubrication modes on part quality during drilling 6061–T6 aluminium alloy. Int. Journal of Machining and Machinability of Materials. 2013;13(2):231-52.

[69] Klocke F, Eisenblätter G. Dry cutting. CIRP Annals-Manufacturing Technology. 1997;46(2):519-26.

[70] Zedan Y, Samuel F, Samuel A, Doty H. Effects of Fe intermetallics on the machinability of heat-treated Al–(7–11)% Si alloys. Journal of Materials Processing Technology. 2010;210(2):245-57.

[71] Dornfeld D. Burr formation, Burr Minimization and Deburring seminar: CRIAQ MANU-409C Automated deburring and part finishing Ecole de technologie superieure (ETS); 2009.

[72] Xi FJ, Liao L, Mohamed R, Liu K. A tripod-based polishing/deburring machine.Smart Devices and Machines for Advanced Manufacturing: Springer; 2008. p. 137-166.

[73] Oliveira JFG, Valente CMO. Monitoring and Control in Abrasive Robotic Deburring Operations. AC ; 2004.7:18.

[74] Lee K, Huang H, Lu S. Adaptive Hybrid Impedance Force Control of Robotic Deburring Processes. Proceedings of the 32nd International Symposium on Robotics. 2001; TC 5-3:1-6.

[75] Asakawa N, Toda K, Takeuchi Y. Automation of chamfering by an industrial robot; for the case of hole on free-curved surface. Robotics and Computer-Integrated Manufacturing. 2002;18(5):379-85.

[76] Bogue R. Finishing robots: a review of technologies and applications. Industrial Robot: An International Journal. 2009;36(1):6-12.

[77] Zhang H, Chen H, Xi N, Zhang G, He J, On-line path generation for robotic deburring of cast aluminum wheels. Intelligent Robots and Systems (IROS), 9-15 Oct. 2006, Beijing, China.

[78] Kazerooni H. Automated robotic deburring using impedance control. Control Systems Magazine, IEEE. 1988;8(1):21-5.

[79] Dornfeld D. Acoustic emission feedback for precision deburring. CIRP Annals-Manufacturing Technology. 1992;41(1):93-6.

[80] Hirabayashi H, Ohwada S, Yoshida I, Miki M. Force-Control Deburring Robots. Society of Manufacturing Engineers (SME), 1987:p. 1-12.

Section 2

Magnesium Alloys

Chapter 6

Thermal Stability, Formability, and Mechanical Properties of a High-Strength Rolled Flame-Resistant Magnesium Alloy

Masafumi Noda, Kunio Funami, Hisashi Mori,
Yoshio Gonda and Kenji Fujino

Additional information is available at the end of the chapter

http://dx.doi.org/10.5772/57467

1. Introduction

Lightweight magnesium (Mg) alloys with excellent shock-absorption properties are being actively adopted for electronic information devices and automotive parts [1]. Mg alloys shows excellent environmental properties because of their light weight, which can lead to improved energy efficiency and hence a reduction in emissions of carbon dioxide [1–2]. For use in structural applications, Mg alloys need to have adequate ductility, thermal stability, and strength. However, Mg alloys often exhibit low ductility, low tensile yield strength, and poor formability as a result of limited slip in their hexagonal close-packed structures [2]. The known ways for effectively improving the mechanical properties and formability of Mg alloys include grain refinement [3–4] and control of the texture of the alloy [4–5]; both these techniques promote prismatic slip and facilitate the creation of large plastic deformations. It is well known that the ignition temperature of Mg alloys is lower than that of other materials for vehicles [6]. However, the ignition temperature of Mg alloys can be markedly raised by the presence of small amounts of calcium (Ca) [6]. Recently, AZX and AMX (X = Ca) alloys have been shown to have higher ignition temperatures than other Mg alloys [6–7]. The effect of adding Ca to improve the flame resistance of Mg alloys has been demonstrated experimentally, as shown in Figure 1. At a temperature at 550 °C, AZ61 Mg alloy ignites, the surface of A6N01 alloy begins to change color, and AZX611 alloy is unaffected.

In general, Mg alloys that contain other elements are known to have greatly enhanced mechanical properties, but their ductility can be maintained only processing the cast metal through extrusion and/or plastic deformation [10–11]. It has been suggested that grain refinement of the Mg phase occurs during extrusion deformation [3,10–12]. The strength and microstructure of extruded Mg–Ca and Mg–Ca–Al(Zn) [6–7,9–13] alloys have been examined, together with the distribution of compounds within these materials [7,12,15]. Most of the plastic deformation processes to which Ca-containing Mg alloys have been subjected are extrusion processes [9–10], and there have been few attempts to examine improvements in the strength of these materials by means of rolling and working processes. It is widely known that the plastic deformation of alloys of Mg with rare earth (RE) metals [8] or Ca [11–12] requires many working cycles and high processing temperatures in comparison with commercial Mg alloys [5,7]. If wrought Mg materials are to be more widely used, it will be important to develop techniques for the production of rolled sheets in addition to extruded materials. It will also be necessary to elucidate the mechanical properties of such materials and to establish a rolling process for Mg alloys that is faster than the current extrusion processes. It has been reported that a rolling process has been carried out at processing temperatures (sample temperature, roll surface temperature, and reheating temperature) above the static recrystallization temperature [6–13]. However, there have been no previous studies on the relationship between the microstructural changes responsible for strength enhancement and rolling processes without reheating or of the maximum reduction in thickness achievable in single-pass rolling.

In this study, we investigated the changes in the microstructure and tensile properties at room temperature produced by controlled rolling of samples of Mg–Al–Mn–Ca cast alloy produced by a twin-roll casting process. The heat resistance, formability, and damping properties of the rolled sheet produced were compared with those of the A6N01 alloy currently used in high-speed rail vehicles [14].

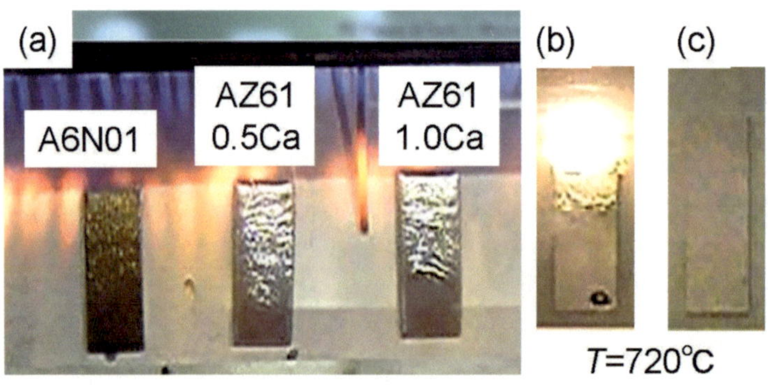

Figure 1. Appearance of samples of A6N01, AZ61+0.5Ca and AZ61+1Ca alloys subjected to flame-resistance testing (a). Ignition temperature of Ca addition magnesium AZX611 alloy (c) higher than AZ61 (b) magnesium alloys.

2. Experimental procedure

The alloy used in this study was twin roll cast (TRC) Mg–10Al–0.2Mn–1Ca (mass%) alloy (AMX1001). The material properties of this alloy were compared with those of samples of

http://dx.doi.org/10.5772/57467

Mg–3Al–1Zn–1Ca (AZX311) alloy and Mg–6Al–1Zn–1Ca (AZX611) alloy. Ingots were prepared in an electric furnace in an atmosphere of argon. The as-received samples of the TRC material measured 100 mm wide by 2000 mm long by 4 mm thick. Samples for rolling, measuring 60 mm wide and 120 mm long, were cut from the as-received materials. The TRC direction and rolling direction were parallel to one another, and the microstructure was observed from the direction perpendicular to the direction of rolling and casting. Rolling was carried out on samples preheated to 100 to 400 °C for 10 min in a furnace. The surface temperature of the rollers was maintained at 250 °C by means of embedded heating elements. The sample was rolled from a thickness of 4 mm to one of 1 mm in several passes (1 mm per pass) without reheating during the rolling process, or by single-pass rolling. The roll diameter was 180 mm and roll speed was set at 5, 10, 15, or 25 m/min. We chose to roll the sheets to a thickness of 1 mm to permit comparison of the mechanical properties of sheets produced by multipass rolling with those of sheets rolled in a single pass. The samples were cooled with water within 5 s of the final rolling pass. The AMX1001 alloy consisted of an α-Mg phase and Al–Ca compounds [13].

Samples for tensile testing with gauge sections 2.5 mm in width and 15 mm in length were machined from samples of the rolled and annealed materials. Tensile tests were carried out at an initial strain rate of 5.0×10^{-4} s^{-1} at room temperature. Samples were annealed at temperatures of between 100 and 400 °C in an electric furnace for various times between 1 and 1000 h, with subsequent cooling in water. The formability of the rolled sheets was investigated by conical cup tests performed temperatures between room temperature and 250 °C at an initial strain rate of 2.7×10^{1} s^{-1}. The microstructures of the rolled and annealed samples were observed by optical microscopy (OM), scanning electron microscopy (SEM), and electron backscattering diffraction (EBSD). The optical and Electron probe micro-analyser (EPMA) maps of as-cast AMX1001 alloy are shown in Figure 2. The initial mean grain size of AMX1001 cast alloy was 53 μm. The brighter areas in Figure 2 correspond to Al_2Ca compounds. The Al_2Ca compounds were present as networks and as coarse agglomerations.

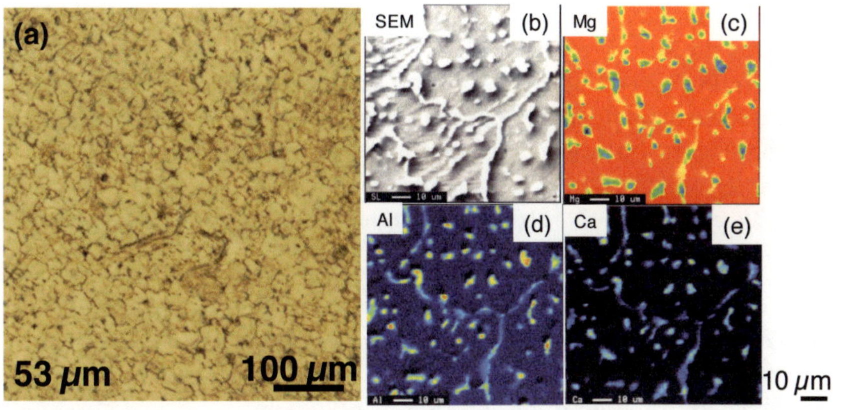

Figure 2. Optical micrograph (a) and EPMA maps (b)-(e) of cast sample of AMX1001 twin roll cast alloy.

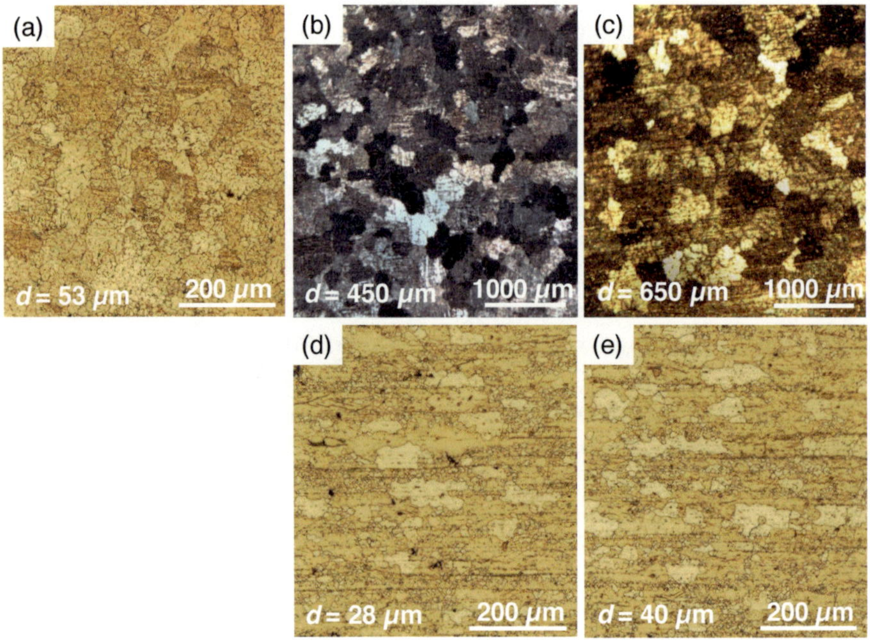

Figure 3. Optical microstructures of twin rolled cast alloy AMX1001 (a), gravity-cast alloys AZX311 (b) and AZX611 (c), and extruded alloys AZX311 (d) and AZX611 (e).

3. Results and discussion

3.1. Structure of the cast materials and the limited reduction in thickness diagram

Rapidly cooled AMX1001 TRC alloy (the as-received material) has a finer grain structure than AZX311 or AZX611 gravity-cast alloys. However, the mean grain size of the AZX311 and AZX611 alloys was similar to that of AMX1001 TRC. The initial optical microstructure of these materials is shown in Figure 3. It is necessary to know the relationship between the sample temperature and the maximal reduction in thickness per pass in order to prepare thin sheets without cracking. Figure 4(a) shows the maximum reduction in thickness per pass in strip pressing of samples of AZX311, AZX611, and AMX1001 cast alloy; optical micrographs of samples of AZX311 and AMX1001 subjected to single-pass rolling at 200 °C are shown in Figure 4(b) and (c).

The maximum reduction in thickness at a deformation temperature of 200 °C for samples of AZX311 and AZX611 cast alloys was 9%, whereas that of AMX1001 cast alloy was 30% [Figure 4(a)]. This shows that AMX1001 TRC alloy has a higher deformability than AZX311 and AZX611 cast alloys. Furthermore, the maximum reduction in thickness reached 50% for AMX1001 TRC alloy when the deformation temperature was raised to 300 °C; the maximum reduction in thickness therefore increases significantly at deformation temperatures between

http://dx.doi.org/10.5772/57467

200 and 300 °C. Optical micrographs recorded after applying a reduction in thickness of 20% to AMX1001 TRC alloy at a sample temperature at 200 °C and a roll surface temperature of 250 °C are shown in Figure 4(b) and (c). It can be seen that, in comparison with the cast material, shear deformation is introduced into the microstructure. As will be discussed later, dynamic recrystallization (DRX) occurs on elevating the sample temperature after the rolling process. As a result of the rolling process, the α-Mg phase becomes finer, Al–Ca compounds are finely crushed, and structural rearrangement occur; furthermore, the Mg phase and the Al–Ca compounds exhibit a lamellar structure.

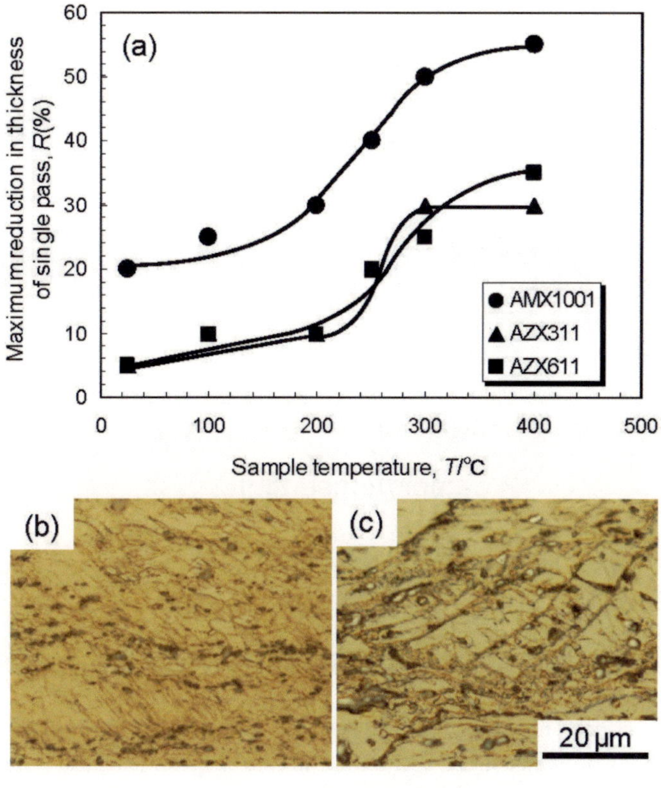

Figure 4. Maximum reduction in thickness curves for AZX311, AZX611, and AMX1001 cast alloy (a) and optical micrographs of AZX311 (b) and AMX1001 (c) alloys rolled at 200 °C in a single pass.

Curves showing the maximum reduction in thickness of extruded AZX311 and AZX611 alloys are presented in Figure 5 [7]. The maximum reduction in thickness of AMX1001 cast material was the same as that of extruded AZX311 and AZX611 alloys. In other words, the TRC material shows a good rollability in comparison with the gravity-cast materials when the mean grain size is less than 100 μm. The TRC process, which involves rapid cooling, therefore has advantages in terms of the rollability of the product, in addition to its greater productivity. By the way, Figure 5 shows that the maximum reduction in thickness of AZX611 alloy decreases when the deformation temperature is increased to more than 400 °C. This decrease is probably the result of growth of grains of α-Mg phase and the proximity of the annealing temperature to the solution-treatment temperature [13]. These observations are consistent with the forging and molding properties of AZ31 alloys [15] and Mg–Zn–Y alloys [8], and also with the fact that AZ31 alloy is a nonaging alloy, whereas AZ61 alloy is an aging alloy. In other words, the roll surface temperature and the reduction in thickness are important factors in strip processing of AMX1001, AZX311, and AZX611 alloys, and a fine dispersion of Al_2Ca compounds occurs as a result of the refinement of the α-Mg phase.

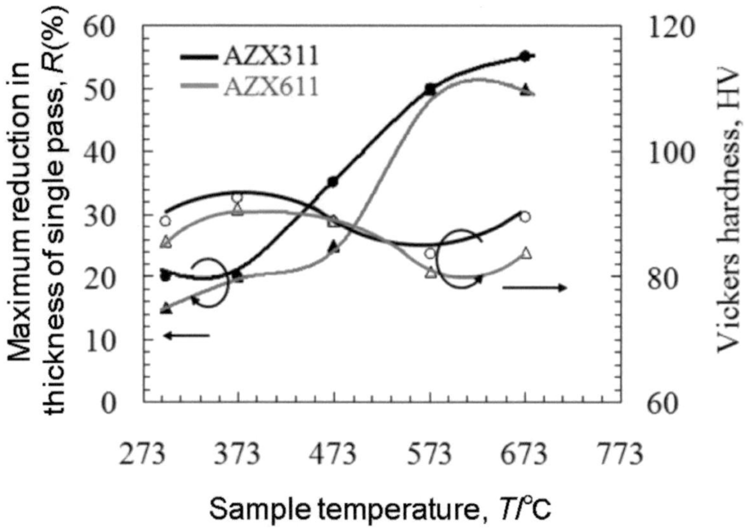

Figure 5. Maximum reduction in thickness curves for AZX311 and AZX611 extruded alloy at 200 °C after a single pass. The results for AMX1001 cast alloys are also shown in this figure [7].

3.2. Effects of rolling conditions on the microstructure of flame-resistant Mg alloy

To investigate the effect of rolling conditions and grain refinement, we used 10-mm-thick samples of AZX611 gravity-cast material with a coarse grain, because increasing the total reduction in thickness eliminated the fine structure of the cast material.

Figure 6(a) shows the tensile properties of the as-cast alloys subjected to rolling at a sample temperature of 200 °C or 400 °C with a roll surface temperature of 250 °C. Samples were rolled from a thickness of 10 to 1 mm in nine passes at 200 °C (the 200 °C rolled sheet) or in a single pass at 400 °C (the 400 °C rolled sheet). The yield stress (YS) of the 200 °C rolled sheet was 340 MPa, its ultimate tensile strength (UTS) was 350 MPa, and its elongation was 4%. The YS and UTS of the 400 °C rolled sheet were 220 and 270 MPa, respectively, and its elongation was 11%. In the single-pass rolling at a sample temperature of 400 °C, we were able to reduce the thickness by 90%, because the sample temperature was close to the solution temperature and the Mg phase and the Al–Ca compounds could be easily deformed. The YS and UTS of the resulting sample were lower than those of the alloy sample rolled at 200 °C, but the elongation was improved. The difference between the temperature of the sample (200 °C) and the roll surface temperature (250 °C) suppresses loss of heat during the rolling process, thereby inducing DRX and increasing plastic deformability. Optical micrographs of the 200 °C and 400 °C rolled materials are shown in Figure 6(b) and (c). The Al–Ca compounds are finely crushed and dispersed during the rolling process, and fine Al–Ca compounds are formed at grain boundaries. It is well known that finely crushing Al–Ca compounds contributes to control of grain growth [7]. A comparison of the textures in optical micrographs of the 200 °C rolled sheet after nine passes and in the 400 °C rolled sheet after a single pass showed that the former contained relatively equiaxial grains, whereas the latter showed a microcrystalline structure between the grain boundaries and a completely uniform texture was not formed. Therefore, strip processing for a few passes under a small load to produce a uniform texture is effective in reducing the total load and increasing the strength of the alloy without reheating.

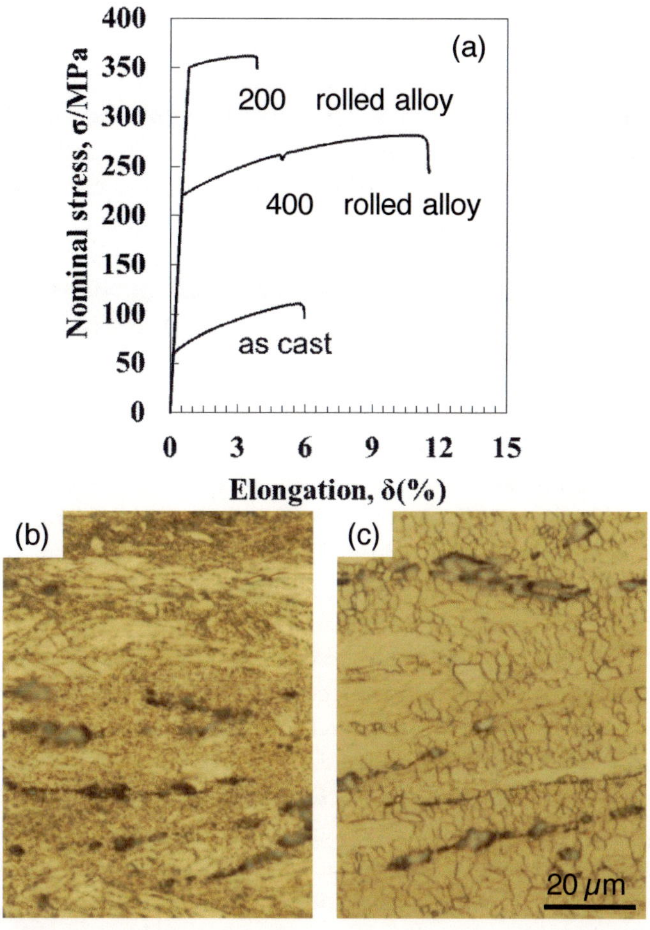

Figure 6. Nominal stress–strain curves for AZX 611 sheet rolled at 200 °C (multipass) or 400 °C (single pass) (a), and optical micrographs of the materials rolled at 200 °C (b) and 400 °C (c).

Figures 7(a–f) show the inverse pole figure (IPF) and pole figure (PF) maps of alloy samples rolled at 200 °C [Figures 7(a–c)] or 400 °C [Figures 7(d–f)]. In the sample rolled at 200 °C, the recrystallized region amounted to 81% of the rolled sheet, and the mean grain size was reduced from 1140 μm to 3.5 μm (the mean grain sizes in the recrystallized and the nonrecrystallized regions were 1.5 μm and 9.4 μm, respectively) (Figure 7a). The sample and roll surface temperatures of 200 °C and 250 °C, respectively, are close to the static and dynamic recrystallization temperatures of Mg alloy [10]. The microstructure was refined by DRX, resulting in a duplex grain structure consisting of partially elongated grains, and shear deformation was observed. As can be seen in Figs. 7(b–c), the intensity of the basal texture of the AZX611 rolled sheet was 10.2; the intensities of the basal textures of the recrystallized and nonrecrystallized regions were 9.8 and 19.6, respectively. As shown in Figures 7(d–f), the sample rolled at 400 °C contained a region with elongated grains and fine grains. The recrystallized region of the rolled sheet accounted for 47% of the total, and the mean grain sizes of the recrystallized and nonrecrystallized regions were 2.9 and 27.8 μm, respectively. As can be seen in Figures 7(a–c), the intensity of the basal texture of the AZX611 rolled sheet was 13.9, and the intensities of the basal textures of the recrystallized and nonrecrystallized regions were 4.5 and 21.4, respectively. The sample temperature of 400 °C is near the grain growth and solution temperature of AZX611 and AZ61 alloys [7,12–13]. The microstructure was refined by DRX, but grain growth occurred immediately after rolling. As a result, the alloy rolled at 400 °C showed a duplex grain structure, the intensity of the basal texture of the nonrecrystallized region was higher than that for the sample rolled at 200 °C, and the area frequency of the nonrecrystallized region was more than 50%. Therefore, if there is a difference in the sample temperature, but no observed significant differences in the mean grain size, provided the nonrecrystallized region is taken into consideration, the grain-refinement mechanism is dependent on the sample temperature and the reduction in thickness.

Next, we focused on the sample temperature after rolling in the nine-pass process that produced high-strength AZX611 rolled sheet. Figure 8(a) shows the results of measurements of the sample temperature after rolling and the mechanical properties for various total reductions in thickness. Figures 8(b–d) show optical micrographs for the several reductions in thickness, as indicated in Figure 8(a). From Figure 8(a), it is apparent that when the total reduction in thickness was less than 60%, the sample temperature after rolling increased slightly as the number of rolling passes increased. On the other hand, the sample temperature after the rolling process did not show any increase for a total reduction in thickness of more than 60%. The strength increased after a total reduction in thickness of 60%, but the elongation was reduced. Figures 8(b–d) show that the dendrite structure was also elongated in the rolling direction after a total reduction in thickness of 40% and that Al–Ca compounds remained in the grain boundary. The existence of shear deformation of the microstructure when the total reduction in thickness reached 60% was confirmed and, at this stage, Al–Ca compounds were beginning to be crushed. When the total reduction in thickness reached 80%, bending of the microstructure occurred as a result of shear deformation and elongation of the grain in the direction parallel to the rolling direction. By performing multipass rolling and introducing shear deformation by the rolling process, we were able to increase the strength of the Mg phase by DRX, while the Al–Ca compounds were crushed. This was possible because the deformation

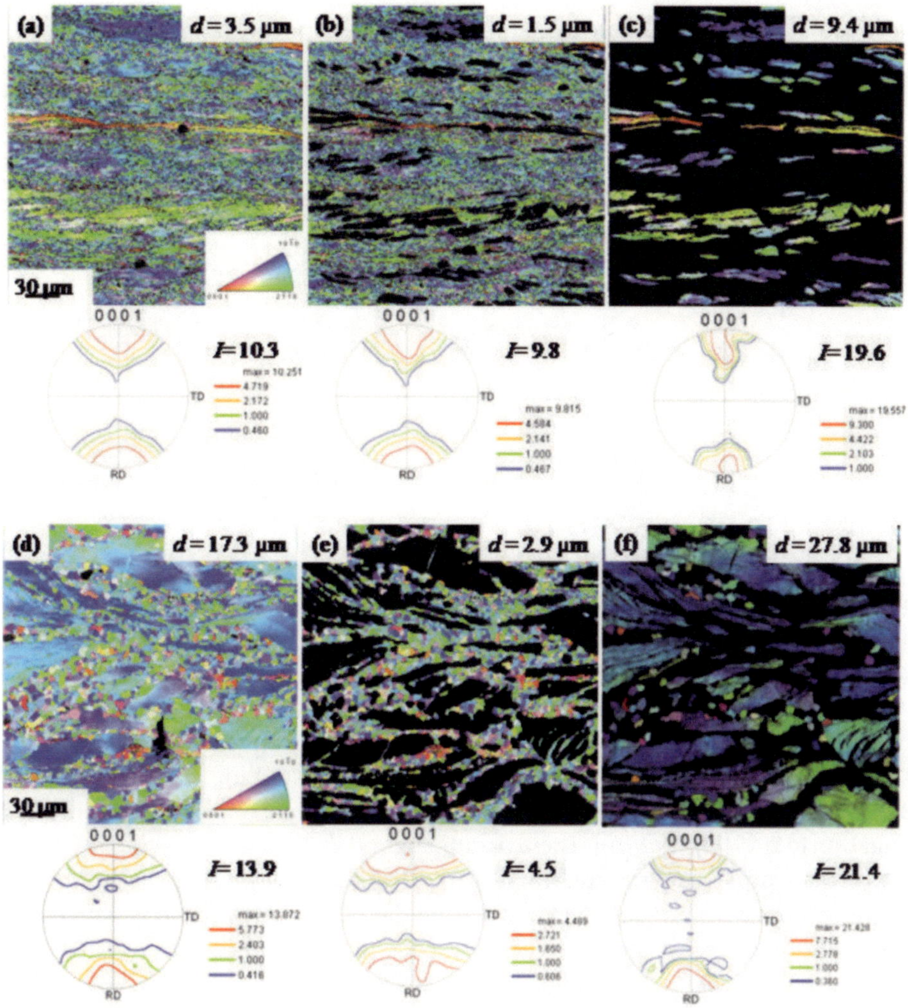

Figure 7. Inverse pole figure and pole figure maps of samples of AZX611 alloy rolled at 200 °C (a)–(c) or 400 °C (d)–(f); (b) and (e) show the recrystallized regions and (c) and (f) show the nonrecrystallized regions.

process induced heating of the alloy during each rolling pass, thereby maintaining the sample at a temperature near its solution temperature. Al–Ca compounds that had been crushed were rearranged in layers in the Mg phase in the direction parallel to the rolling direction.

3.3. Production of high-strength Mg–10Al–0.2Mn–1Ca rolled sheet by a rolling process

From Figures 2 and 3, the initial grain size of AMX1001 TRC alloy was lower, by a ratio of 1:20, than the mean grain size of AZX311 and AZX611 gravity-cast alloy, so that improved rollability would be expected. Therefore, a sample of AMX1001 was rolled from a thickness of 4 mm to 1 mm in three passes. Figures 9(a–c) show the relationship between the mechanical properties and various rolling conditions for AZX1001 alloy. Figure 9(a) shows that, as a result of the

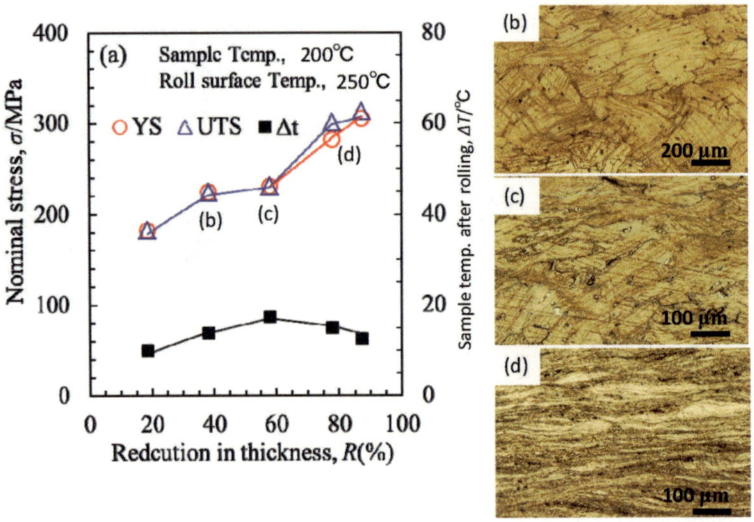

Figure 8. Relationship between sample temperature after the rolling process and the mechanical properties of AZX611 (a); optical micrographs for samples of AZX611 reduced in thickness by 40% (b), 60% (c), and 80% (d).

multipass rolling process, the YS decreased slightly from 390 to 340 MPa and UTS also decreased slightly from 410 to 380 MPa, probably due to the increase in sample temperature from 100 to 350 °C, whereas the elongation improved from 3.5 to 8.5%. Additionally, it was clear that the YS and UTS of samples subjected to single-pass rolling were lower than those of samples subjected to multipass rolling process, but the single-pass rolling markedly improved the elongation to 20% [Figure 9(b)]. These results suggest that the elongation of rolled materials depends on the sample temperature, whereas the YS and UTS depend on the number of rolling passes. The multipass rolling regime that we used did not involve reheating between individual passes, and we suggest that recovery and DRX do not act effectively during the initial passes of the rolling process. From Figures 7 and 8, it is clear, however, that recovery and DRX do act effectively when the reduction in thickness reaches more than 70%. Figure 9(c) shows that the YS and UTS decrease with increasing rolling speed for a total reduction in thickness of 80%, even with a multipass rolling process. This is because there is an increase in the sample temperature after rolling, as in the case of extrusion [8]. It is possible to combine single- and multipass rolling processes to suit particular applications; the rolling speed and sample temperature are the important factors in the rolling process. In addition, a total reduction ratio of 60% or more is necessary to produce high-strength rolled material.

Figures 10(a–c) show optical micrographs of the as-cast sample and samples of rolled sheet subjected to rolling for three passes, together with the corresponding IPF and PF maps for a sheet rolled at a speed of 10 m/min at a sample temperature of 200 °C and a roll surface temperature of 250 °C. The microstructure of the cast ingot consisted of a coarse grain structure with Al_2Ca and Al compounds. After the three-pass rolling process, the recrystallized region of the rolled sheet accounted for 70% of the total, and the grain size was reduced from 53 to 3.8 μm. The rolling temperature of 200 °C is close to the recrystallization temperature of both

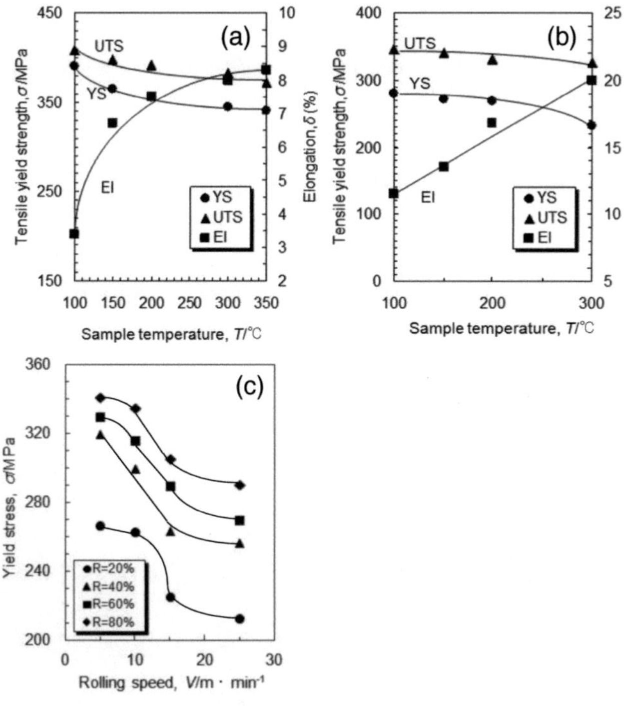

Figure 9. Relationship between the mechanical properties and rolling conditions for a multipass process (a) and a single-pass process (b) performed at various sample temperatures. Relationship between mechanical properties and rolling speed at a sample temperature of 200 °C and roll surface temperature at 250 °C for a reduction in thickness of 80% (c).

AZX and AZ series alloys [11–12,15]. The microstructure was refined by DRX to form a duplex grain structure that was partially elongated, and shear deformation was observed into the elongated grain. To increase the extent of the recrystallized region, an increase in the rolling temperature or a greater total reduction thickness was required, but the strength then tended to decrease because of the influence of grain growth. The Al–Ca compounds were finely crushed and dispersed during the rolling process, and therefore these compounds contributed to control of the grain growth that would otherwise result from an increase in the rolling temperature. As can be seen in Figure 10(c), the intensity of the basal texture of the AMX1001 rolled sheet was 8.2. This basal texture was lower than those of AZ31 alloy samples subjected a single rolling process at 200 or 400 °C, until the rolling reduction reached 86%, at which point the intensities were 7 and 5, respectively [16]. Figure 11 shows the tensile properties of the as-cast and rolled (single- and three-pass schedules) alloys for a roll surface temperature of 250 °C and a rolling speed of 10 m/min; multipass rolling was performed at 100 °C, and single-pass rolling was performed at a sample temperature of 400 °C. The YS and UTS of the sheet subjected to three-pass rolling were 380 and 400 MPa, respectively, and the elongation was 8%. A sheet rolled from a thickness of 4 mm to one of 1 mm in a single pass showed a UTS of 320 MPa, a YS of 220 MPa, and an elongation of 15%. The failure to improve the strength of the material by the three-pass rolling process was the result of the heat generated during the metal-forming process. In fabricating the 1-mm-thick rolled sheet by the three-pass rolling process, grain refinement of the Mg phase and crushing of the Al-Ca compounds occurred. The difference between the temperature of the sample and that of the roll suppresses heat removal during the rolling process, thereby inducing DRX and increasing deformability.

If no reheating is incorporated into the rolling process, the sample and roll-surface temperatures are close to the recrystallization temperature of the Mg alloy [7,15]. As a result, it is possible to induce DRX through the increase in sample temperature generated by repeated rolling. As a result of the temperature difference between the sample and roll surface, the sample temperature approaches the roll surface temperature more closely as the number of passes increases, making this an important factor.

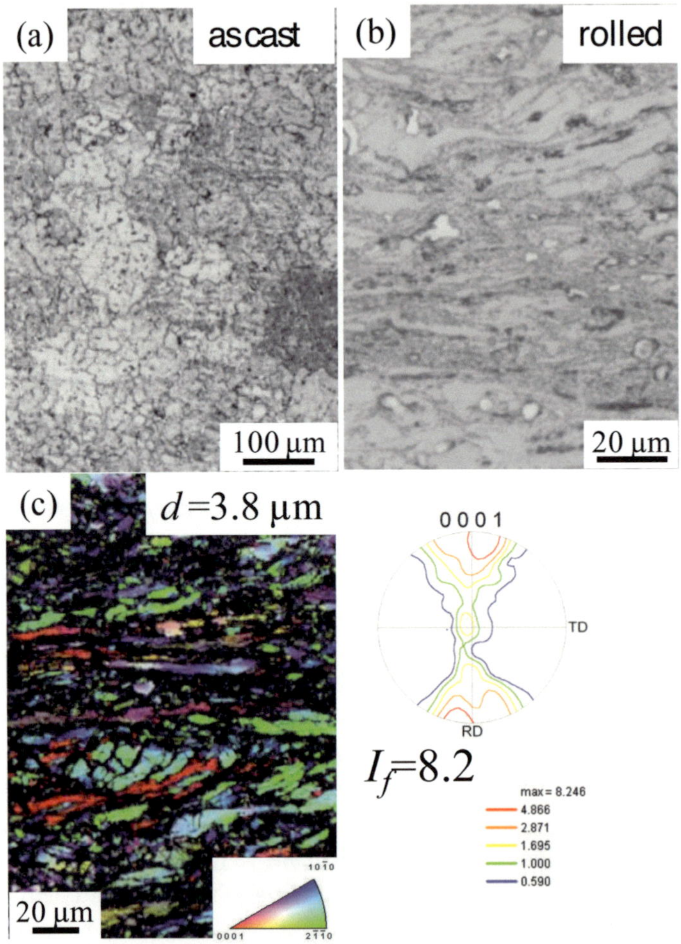

Figure 10. Optical micrographs (a)-(b) and inverse pole figure map (c) of as rolled AMX1001 alloy. The thickness of rolled materials was 1 mm (a total reduction in thickness of 75%).

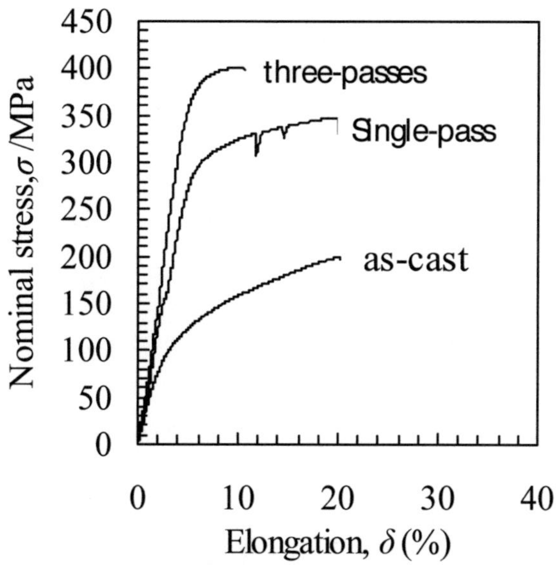

Figure 11. Nominal stress–strain curves for as-rolled AZX1001 alloy samples subjected to multipass rolling at 200 °C or single pass rolling at 400 °C.

3.4. Annealing conditions and the thermal stability of AMX1001 rolled sheet

Figure 12(a) shows the relationship between the mechanical properties and the annealing temperature for rolled sheets of AMX1001. Figure 12(b)-(e) shows optical micrographs of rolled sheets of AMX1001 annealed at various temperatures between 150 and 400 °C. The YS and UTS decreased by about 40 MPa and the elongation increased to 11% when the annealing temperature was raised from 150 °C to 200 °C. A further increase in the annealing temperature to 300 °C resulted in a significant decrease in the YS to 260 MPa and in the UTS to 310 MPa, whereas differences in the YS, UTS, and elongation for annealing temperatures between 300 and 400 °C were minimal. An AMX1001 rolled sheet subjected to annealing at 200 °C for 1 h showed no significant reduction in strength or tensile properties, which were are similar to those of high-strength Mg alloy. In other words, the AMX1001 alloy showed excellent thermal stability as a result of the addition of a small amount of Ca [10–11]. The addition of Al, however, did not appear to have any effect on the thermal stability. The changes in the strength and elongation of the AMX1001 rolled sheet suggest that static recrystallization occurs at 200 to 250 °C [Figures 12 (a) and (b)-(c)]. The deterioration in mechanical properties can be effectively controlled by suppressing grain growth. Figure 13 shows that Al–Ca compounds form along the grain boundaries after annealing, but that some compounds form in the grain. Focusing on the grain size, fine grains are seen in the material annealed at 300 °C, whereas 18-μm grains are seen in the material annealed at 350 °C; furthermore, there is also a decrease in the formation of Al–Ca compounds. As the annealing temperature was increased, the Al–Ca compounds were able to control the grain growth, and grain coarsening occurred rapidly at temperatures close to the solution temperature. The other elements in the alloy are considered to be partially soluble at 400 °C, and the formation of Al_2Ca compounds has been reported to be effective in improving the ductility of alloys [13]. The alloy in this work, in which Al_2Ca compounds were formed as a result of the addition of 1 mass% of Ca, is considered to retain its ductility while showing a greater strength and larger elongation than other Mg alloys.

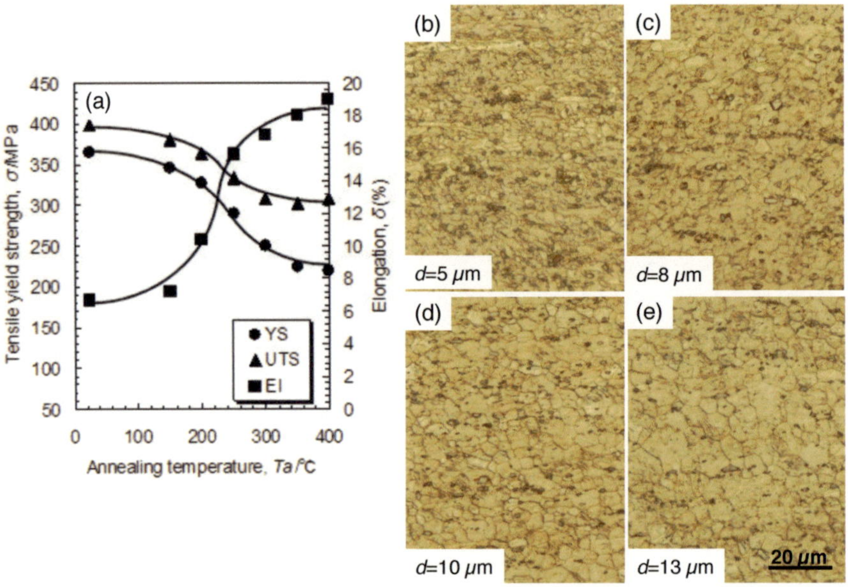

Figure 12. Tensile properties (a) and optical micrographs (b)-(e) of samples of AMX1001 alloys annealed at 200°C(b), 250°C(c), 300°C(d) and 400°C(e) for 1 h.

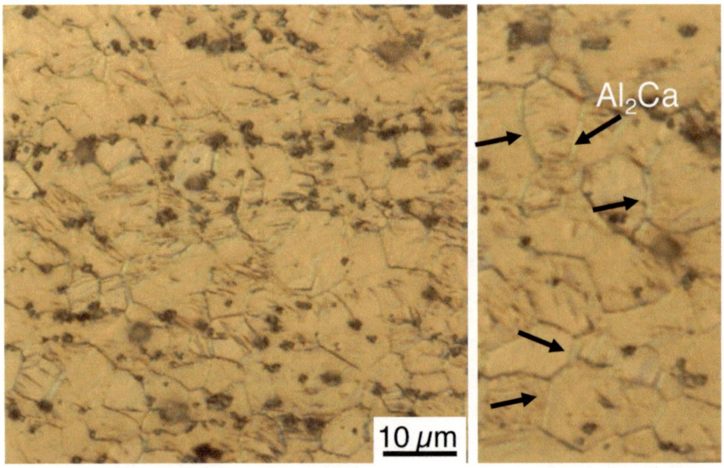

Figure 13. Optical micrographs of the typical dispersion state of Al–Ca compounds in the alloy after annealing at 400 °C for 1 h.

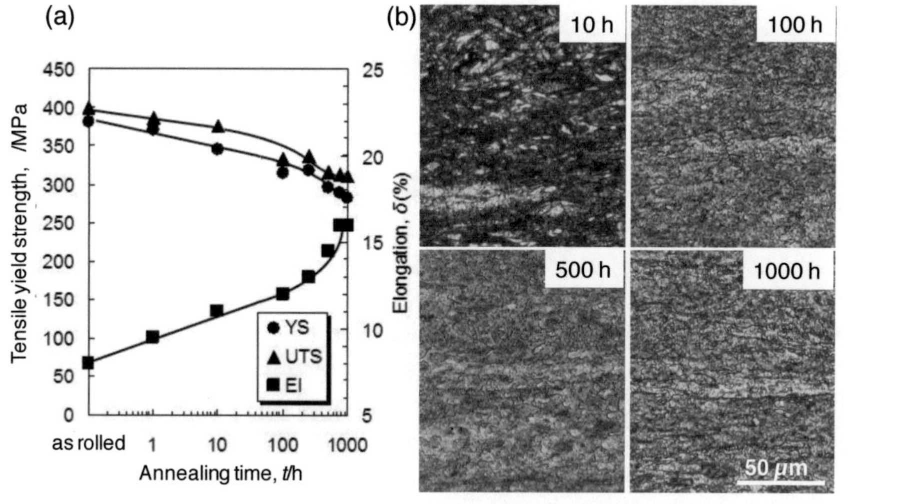

Figure 14. Relationship between the annealing time and the mechanical properties at room temperature of rolled samples of AMX1001 alloy annealed at 200 °C (a). Optical micrographs of rolled samples of AMX1001 alloy annealed at 200 °C for various holding times (b).

The rolled samples of AMX1001 alloy did not show any marked loss of strength or changes in microstructure on annealing at 200 °C for 1 h. We therefore extended the annealing time to 1000 h to test the thermal stability of the AMX1001 rolled sheet. Figure 14(a) shows the YS and elongation for a sample of rolled AMX1001 alloy annealed at 200 °C for various times up to 1000 h, as tested at room temperature. The YS and UTS decreased gradually with increasing annealing time, whereas the elongation markedly improved. The tensile properties of samples annealed at 200 °C for 1000 h did not depend on the annealing time. Figure 14(b) shows optical micrographs of samples annealed at 200 °C for various annealing time. Figure 14(a) shows that when the annealing temperature was maintained at 200 °C for 1000 h, even though the YS was reduced from 390 to 280 MPa, the elongation improved from 8 to 22%. Although the α-Mg phase grew from 4 to 10 μm, and a lamellar microstructure in which the Al_2Ca compounds was finely dispersed in the α-Mg phase was formed, no substantial changes in the microstructure were observed, even after annealing for 1000 h. An examination of the optical micrographs in Figure 14(b) shows that when the sample was annealed for 1000 h, the lamellar microstructures of the α-Mg and Al_2Ca compounds were the same as those observed before annealing. In other words, degradation of the mechanical properties of the AMX1001 rolled sheet after annealing at 200 °C is due to static recovery of the alloy. With regard to the reinforcing factor of this material, Al_2Ca compounds control the grain growth of α-Mg phase even after annealing at 200 °C for 1000 h. The formation of Al_2Ca compounds by adding 1 mass% of Ca is therefore an effective way of increasing the heat resistance of the alloy. The YS and UTS of rolled AMX1001 alloy were lower than those of Mg–Zn–Y extruded alloy [8] after heat treatment at 200 °C for 1000 h; however, AMX1001 alloy can be fabricated into thin rolled sheet at rolling temperature below 200 °C with a small number of passes. In the case of

Mg alloys, it is important that they retain a high strength and a high ductility if they are to be used as industrial materials, and the Ca-containing Mg alloy AMX1001 is a material that possesses such properties.

3.5. Formability and damping property of flame-resistant magnesium alloy

The formability of AMX1001 high-strength rolled sheets was examined by means of a conical cup tests performed at room temperature to 250 °C and an initial strain rate of 2.7×10^{-1} s^{-1}. Specimens measuring 36 mm in diameter were cut from AMX1001 rolled sheet and subjected to conical-cup tests, the results of which are shown in the Figure 15. For comparison, Figure 15 also shows the conical cup value for AZ61 Mg rolled sheet and for high-strength rolled sheet 6N01 Al alloy (Al–0.58Mg–0.6Si mass% alloy) [14], which has a YS of 480 MPa, a UTS of 497 MPa, and an elongation of 8%, and is used in high-speed rail vehicles. The conical cup value of AMX1001 rolled sheet was 27 at 150 °C or above; this value was not significantly improved by increasing the testing temperature to 200 °C. At test temperatures of up to 100 °C, the 6N01 Al rolled sheet showed a better formability than the AZ61 and AMX1001 Mg rolled sheet; however, at a test temperature of 150 °C, the conical cup values for the Mg and Al alloys were very similar. The conical cup value for the AMX1001 rolled sheet was therefore excellent at test temperatures of 150 °C or more. At high testing temperatures, samples of AZ61 and AMX1001 alloy produced by low-load processing showed higher conical cup values than did high-strength 6N01 Al rolled sheet. AMX1001 rolled sheet fabricated by the rolling process described in this study therefore showed better formability than fine-grained 6N01 Al rolled sheet. Figure 16 shows optical micrographs and IPF maps of AMX1001 rolled sheet after conical cup testing at 150 °C and 200 °C. The observation area was 500 μm from the fracture tips of the crown part. The microstructure after conical cup testing at 150 °C showed an elongated grain in comparison with the as-rolled microstructure shown in Figure 10. Along with the improvement in formability demonstrated by the conical-cup test, equiaxial grains were formed in the microstructure at a testing temperature of 200 °C; additionally, the nonrecrystallized region changed to a recrystallized region and a fine grain structure formed with without elongation when the temperature of deformation was high. In other words, to improve the formability of the Mg alloy, it is necessary to select an appropriate temperature and to make use of DRX during plastic deformation. The IPF maps shown in Figure 16 show that at a testing temperature of 200 °C, crystal orientation was random and the texture of the sample after the conical-cup test was weak in comparison with that observed after testing at 150 °C. We found that at a testing temperature of 200 °C, DRX occurred during plastic deformation.

It is well known that Mg alloys have excellent vibration-absorbing properties. The vibration properties of Mg alloys are often reported [17–18], but few comparisons have been made with steel or Al alloys. We examined the damping properties of samples of rolled and/or extruded steel, Al alloys, and Mg alloys by analyzing the waveforms produced after a displacement of 0.45 mm by performing a cantilever vibration test. The cantilever test specimens measured 18 mm wide, 200 mm long, and 1 mm thick. We investigated the damping ratio at a strain amplitude of 2×10^{-4} s^{-1} and a displacement of 0.45 mm at the tip of the cantilever. The damping properties of steel (SUS 304), aluminum alloys (A7075, A5083, A6063, A6N01), and magnesium

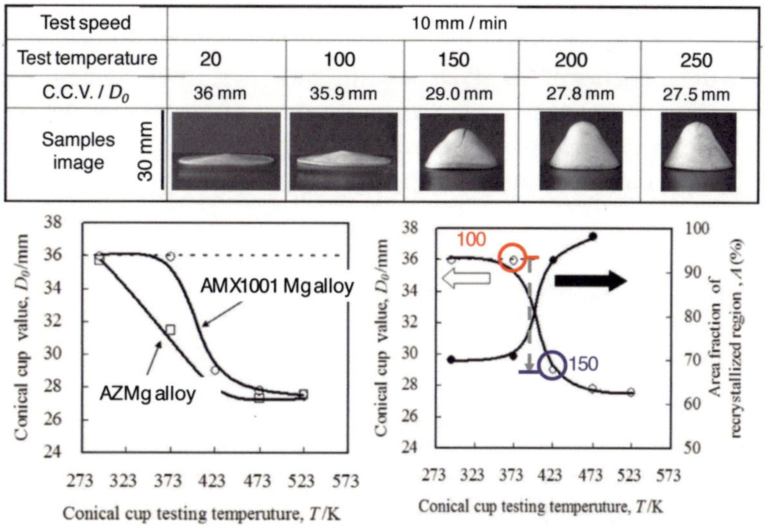

Figure 15. Appearance of conical cup sample and the relationship between the conical cup value and the testing temperature. The conical cup test was performed at various temperatures at an initial strain rate of 2.7×10^{-1} s^{-1}.

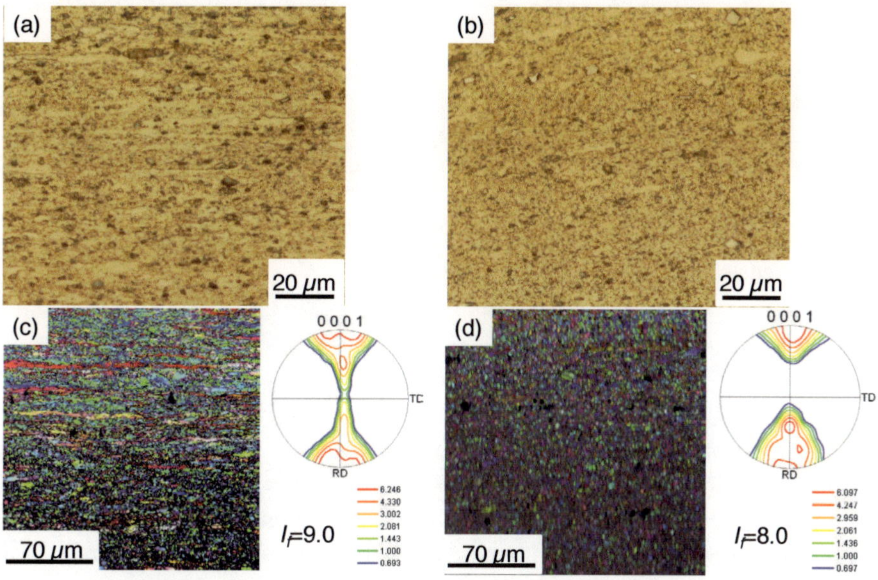

Figure 16. Optical micrographs (a), (b) and inverse pole figure maps (c), (d) of AMX 1001 rolled sheet after conical cup testing at 150 °C (a), (c) and 200 °C (b), (d) at an initial strain rate of 2.7×10^{-1} s^{-1}.

alloys (AZ series, Mg-RE, AZX, and AMX) are shown in Figure 17. From Figure 17, the damping properties improved in the order Mg alloys ≥ Al alloys ≥ steel. The damping ratios of the Mg alloys were dependent on the type of alloy, as in the case of the mechanical properties, and they were affected by the nature of the elements forming the alloy. However, the variation in the percentage of damping ratio on changing of elemental content was small. The material type and the damping ratio showed a linear relationship. In the case of Mg alloys, alloys in the region between the AZ series of Mg alloys and Mg–RE alloys, where flame-resistant Mg alloys occur, had damping properties that were inferior to those of commercial AZ-series Mg alloys, showing that these properties are weakened by the addition of Ca. However, no effect of the addition of Al on the damping ratio could be identified.

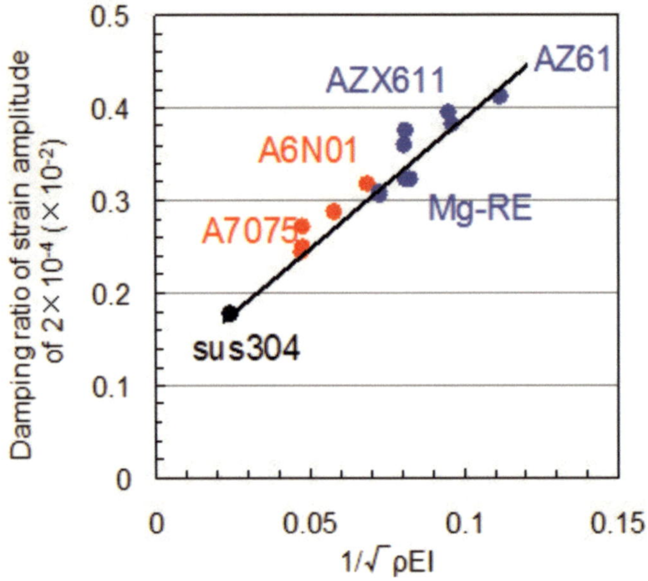

Figure 17. Damping properties of various rolled or extruded materials.

4. Summary

We investigated various properties of flame-resistant Mg alloys. By subjecting TRC materials to a total reduction in thickness of up to 75% by multipass rolling without reheating, we produced a rolled sheet with a tensile strength of 400 MPa and an elongation of 8%. During the multipass rolling process, grain refinement occurred as a result of dynamic recrystallization of the Mg phase and crushing of Al_2Ca compounds. A study of the heat-treatment properties of AMX1001 high-strength rolled sheet found that the yield strength was reduced from 330 to 250 MPa on heating at between 200 and 300 °C for 1 h, whereas the elongation improved from 8 to 17%, suggesting that by static recrystallization had occurred. When we investigated the mechanical properties after heating the material at 200 °C for 1000 h, the tensile strength reached 280 MPa and the elongation improved to 22%. In other words, AMX1001 rolled sheet have an excellent thermal stability. Furthermore, the conical cup value for AMX1001 rolled sheet was maximal at a test temperature of 150 °C or more, and this value was superior to that

http://dx.doi.org/10.5772/57467

of fine-grained rolled A6N01 alloy. A study of the damping properties of various alloys showed that they improved in the order steel ≤ aluminum alloys ≤ magnesium alloys. Overall, the properties of high-strength AMX1001 rolled sheet are superior to those of fine-grained aluminum alloys. In particular, this Mg alloy shows excellent thermal stability, damping properties and formability.

Author details

Masafumi Noda[1*], Kunio Funami[1], Hisashi Mori[2], Yoshio Gonda[3] and Kenji Fujino[4]

*Address all correspondence to: mk-noda@s7.dion.ne.jp

1 Department of Mechanical Science and Engineering, Chiba Institute of Technology, Chiba, Japan

2 Materials Technical Division, Railway Technical Research Institute, Kokubunji Tokyo, Japan

3 Research and Development, Magnesium Division, Gonda Metal Industry Co., Ltd., Kanagawa, Japan

4 Research and Development Center of East Japan Railway Company Group, Saitama, Japan

References

[1] Luo AA, Recent Magnesium Alloy Development for Automotive Powertrain Applications. Materials Science Forum 2003; 419–422 57–66.

[2] Yoo MH, Slip, Twinning, and Fracture in Hexagonal Close-Packed Metals. Metallurgical Transactions A 1981; 12(3) 409–418.

[3] Ion SE, Humphreys FJ, White SH, Dynamic Recrystallization and the Development of Microstructure During the High Temperature Deformation of Magnesium. Acta Metallurgica 1982; 30(10) 1909–1919.

[4] Myshlyaev MM, McQueen HJ, Mwembela A, Konopleva E, Twinning, Dynamic Recovery and Recrystallization in Hot Worked Mg–Al–Zn Alloy. Materials Science and Engineering A 2002; 337(1–2) 121–133.

[5] Watanabe H, Ishikawa K, Effect of Texture on High Temperature Deformation Behavior at High Strain Rates in a Mg–3Al–1Zn Alloy. Materials Science and Engineering A 2009; 523(1–2) 304–311.

[6] Akiyama S, Ueno H, Sakamoto M, Hirai H, Kitahara A, Development of Noncombustible Magnesium Alloys. Materia Japan 2000; 39(1) 72–74 (in Japanese). DOI 10.2320/materia.39.72.

[7] Noda M, Sakai N, Funami K, Mori H, Fujino K, High Strength and Grain Refinement of Mg–3Al–1Zn–1Ca Alloy by Rolling. Journal of the Japan Society for Technology of Plasticity 2013; 54(625) 143–147; DOI 10.9773/sosei.54.143.

[8] Noda M, Kawamura Y, Mayama T, Funami K, Thermal Stability and Mechanical Properties of Extruded Mg–Zn–Y Alloys with a Long-Period Stacking Order Phase and Plastic Deformation: In: Monteiro W A (ed.): New Features on Magnesium Alloys. Rijeka: InTech; 2012. p101–118. Available from http://www.intechopen.com/books/new-features-on-magnesium-alloys (accessed 26 Aug 2012).

[9] Zhou N, Zhang Z, Dong J, Jin L, Ding W, High Ductility of a Mg–Y–Ca Alloy via Extrusion. Materials Science and Engineering A 2013; 560 103–110.

[10] Xu SW, K Oh-ishi, Sunohara H, Kamado S, Extruded Mg–Zn–Ca–Mn Alloys With Low Yield Anisotropy. Materials Science and Engineering A 2012; 558 356–365.

[11] Yim CD, You BS, Lee JS, Kim WC, Optimization of Hot Rolling Process of Gravity Cast AZ31–*x*Ca (x = 0–2.0 mass%) Alloys. Materials Transactions 2004; 45(10) 3018–3022.

[12] Watanabe H, Yamaguchi M, Takigawa Y, Higashi K, Optimization of Rolling Conditions in Mg–Al–Ca Alloy Containing Insoluble Second Phase Particles. Materials Transactions 2008; 49(6) 1262–1269.

[13] Aljarrah M, Medraji M, Wang X, Essadipi E, Muntasar A, Denes G, Experimental Investigation of the Mg–Al–Ca system. Journal of Alloys and Compounds 2007; 436(1–2) 131–141.

[14] Kusuhara H, Noguchi M, Noda M, Mori H, Funami K, Effect of Fine Grain on Mechanical Properties of A6N01 Alloy. Materials Science Forum 2013; 753 501–504.

[15] Matsumoto R, Osakada K, Effect of Heat Treatment on Forgeability of AZ31 Magnesium Alloy. Journal of Japan Institute of Light Metals 2007; 57(7) 274–279.

[16] Chino Y, Mabuchi M, Kishihara R, Hosokawa H, Yamada Y, Wen C, Shimojima K, Iwasaki H, Mechanical Properties and Press formability at Room Temperature of AZ31 Mg Alloy Processed by Single Roller Drive Rolling. Materials Transactions 2002; 43(10) 2554–2560.

[17] Zhou AG, Basu S, Barsoum MW, Kinking Nonlinear Elasticity, Damping and Microyielding of Hexagonal Close-Packed Metals. Acta Materialia 2008; 56(1) 60–67.

[18] Drits ME, Rokhlin LL, Sheredin VV, Shulga YN, Magnesium Alloys with High Damping Capacity. Metal Science and Heat Treatment 1970; 12(11) 939–941.

Chapter 7

Different Performance of Factors Affecting the Estimation of the Corrosion Rate in Magnesium Alloys by Implementation of the Common Methods for Electrochemical Measurements

Alejandro Samaniego and Sebastián Feliu (Jr)

Additional information is available at the end of the chapter

http://dx.doi.org/10.5772/58370

1. Introduction

The attractive properties of magnesium-based alloys make them materials of considerable interest, but their insufficient corrosion resistance is sometimes an impediment to their widespread use. This explains why much current research is focused on better understanding and improving the corrosion behaviour of these alloys.

Such research often relies on the results of immersion tests in aqueous chloride solutions, which have generally proven to be highly appropriate for a quick characterisation and classification of the corrosion resistance of the different magnesium alloys. Corrosion tests in chloride-containing media may, however, be subject to a serious problem as has been noted by several authors [1-3], in that important discrepancies are sometimes seen between the magnitude of the experimental values and expected values based on available information on the typical behaviour of these alloys. In this respect it is a well known fact that much lower electrochemical corrosion rate values may be obtained with magnesium alloys immersed in some corrosive media than the real values determined gravimetrically. In the literature, this behaviour has tentatively been associated with several phenomena such as material disintegration, the participation of monovalent Mg^{+}ions in the corrosion process, and anomalous chemical dissolution, although the true causes of such discrepancies remains to be fully clarified.

The purpose of this chapter is to report part of the results of a study carried out on Mg-Al alloys which provides new information on the occasional occurrence of anomalous corrosion rates in the study of these alloys. The ultimate aim is to contribute to obtain a best picture of such phenomenon, and to improve from the design of the tests the reliability of the experimental results, removing factors potentially responsible for the anomalous behaviour.

Specifically, this chapter addresses the effect on the electrochemical measurements of the corrosion rate of AZ31 and AZ61 alloys of the: (a) configuration of the exposed surface area of the specimens; (b) surface conditions at the start of the test; and (c) measurement time using the two aforementioned commercial magnesium-aluminium alloys. Special attention has been paid to the possible repercussion of the confinement of the exposed area by means of a circular plastic gasket, as well as to the presence of cut edges on the square coupon specimens entirely exposed to the aggressive solution. Depending on the special circumstances of each case, examination of the corroded surfaces may show substantially varied morphologies, which include uniform corrosion, filiform corrosion, and localized attack with the formation of pits and cavities.

2. Experimental

Specimens. The experimental specimens used in the study represent the effect of the following variables: two types of Mg-Al alloys, two surface states, and three different surface areas exposed to the corrosive solution. As will be seen below, all of these variables may have affected the corrosion behavior of the specimens immersed in 0.6 M NaCl in some way.

The tested alloy types were AZ31 and AZ61, whose nominal compositions (wt.%) are listed in Table 1. The tests included surfaces in the as-received condition and polished surfaces to a mirror finish. The geometric surfaces areas exposed to the solution were: 0.5 cm^2, 0.8 $cm^2$and 9.0 cm^2, defined by the circular contour of a plastic polymer gasket.

Alloy	Al	Zn	Mn	Si	Fe	Ca	Mg
AZ31	3.1	0.73	0.25	0.02	0.005	0.0014	Bal.
AZ61	6.2	0.74	0.23	0.04	0.004	0.0013	Bal.

Table 1. Composition of alloys

3. Results and discussion

3.1. Corrosion rates

The graph in Fig.1 plots the normalized corrosion rates (NCR), calculated by dividing the experimentally determined corrosion rate for each specimen by the median value of the set of specimens to which the considered specimen belongs (Tables 2-5). As can be seen, there is some variability in the points in relation to the median value indicated by the line NCR=1. The greatest degree of deviation is obviously shown by the outliers located well above the level NCR=3, with values that multiply the median value by as much as 5-20 times.

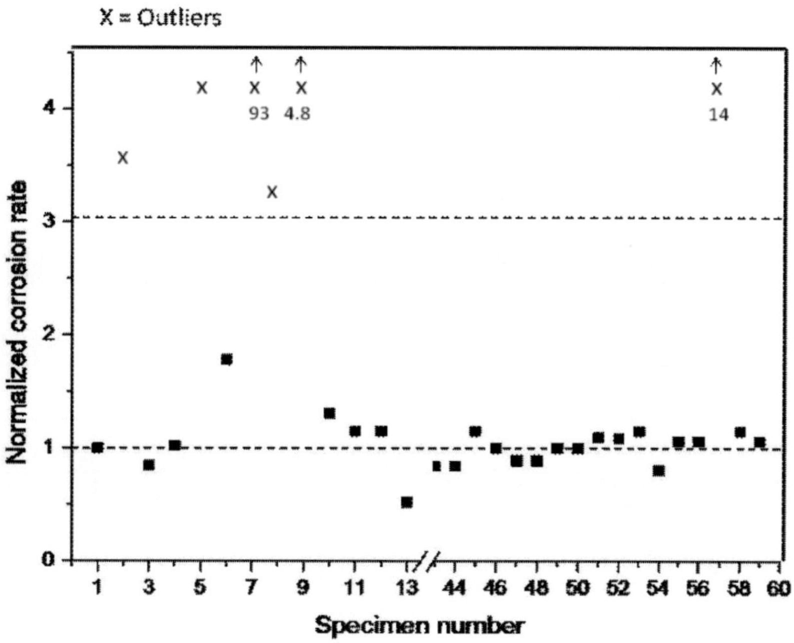

Figure 1. Ratio between gravimetric (GRAV.) and electrochemical (ELEC.) data. The points marked with the letter A correspond to specimens that showed anomalous dissolution rates in weight loss measurements.

Corrosion rate (mg/cm²/d)	Normalized corrosion rate	Specimen No.	Days immersion	Exposed surface (cm)²	Measurement method
0.26	0.41	18	10	10.5	W
0.31	0.48	17	10	10.5	W
0.38	0.59	70	10	10.4	H

Corrosion rate (mg/cm²/d)	Normalized corrosion rate	Specimen No.	Days immersion	Exposed surface (cm)²	Measurement method
0.38	0.59	69	10	10.4	H
0.49	0.77	62	7	10.4	H
0.50	0.78	88	4	10.4	H
0.50	0.78	87	4	10.4	H
0.51	0.80	61	7	10.4	H
0.54	0.84	43	4	10.4	H
0.54	0.84	44	4	10.4	H
0.57	0.89	47	4	10.4	H
0.57	0.89	48	4	10.4	H
0.63	0.98	78	4	0.8	H
0.64	1.00	1	10	0.4	W
0.68	1.06	55	4	10.4	H
0.68	1.06	56	4	10.4	H
0.84	1.31	10	3	9	W
0.93	1.45	86	4	6	H
1.16	1.81	36	1	10.4	H
1.21	1.89	35	1	10.4	H
1.70	2.66	85	4	6	H
2.30	3.59	2	10	0.4	W
2.71	4.23	77	4	0.8	H
3.05	4.77	9	3	9	W
7.55	11.8	72	6	0	H
9.22	14.4	57	4	0	H
15.7	24.5	71	3	0	H

Table 2. Average corrosion rates arranged in ascending order. Specimens of AZ31 alloy in the as-received surface condition (Measurements: W, by weight loss; H, by hydrogen evolution).

http://dx.doi.org/10.5772/58370

Corrosion rate ($mg/dm^2/d$)	Normalized corrosion rate	Specimen No	Days immersion	Exposed surface (cm^2)	Measurement method
0.20	0.32	19	10	10.5	W
0.21	0.34	20	10	10.5	W
0.53	0.85	3	10	0.4	W
0.53	0.85	64	7	10.4	H
0.54	0.87	63	7	10.4	H
0.55	0.89	91	4	10.4	H
0.55	0.89	92	4	10.4	H
0.56	0.90	90	4	6	H
0.62	1.00	49	4	10.4	H
0.62	1.00	50	4	10.4	H
0.63	1.02	4	10	0.4	W
0.63	1.02	89	4	6	H
0.68	1.10	80	4	0.8	H
0.71	1.15	11	3	9	W
0.71	1.15	12	3	9	W
1.31	2.11	37	1	10.4	H
1.40	2.26	38	1	10.4	H
10.90	17.6	79	4	0.8	H

Table 3. Average corrosion rates arranged in ascending order. Specimens of the AZ31 alloy in the as-polished surface condition. (measurements: W, by weight loss; H, by hydrogen evolution).

Corrosion rate ($mg/cm^2/d$)	Normalized corrosion rate	Specimen No	Days immersion	Exposed surface (cm^2)	Measurement method
0.54	0.18	82	4	0.8	H
0.88	0.29	14	3	9	W
1.03	0.33	21	10	10.5	W
1.28	0.42	94	4	6	H

Corrosion rate ($mg/cm^2/d$)	Normalized corrosion rate	Specimen No	Days immersion	Exposed surface (cm^2)	Measurement method
1.30	0.42	26	3	10.4	W
1.35	0.44	22	10	10.5	W
1.49	0.48	60	4	0.84	H
1.51	0.49	76	6	0.8	H
1.61	0.52	13	3	9	W
1.80	0.58	25	3	10.4	W
2.13	0.69	29	3	10.4	W
2.75	0.89	96	4	10.4	H
2.82	0.92	73	10	10.4	H
2.82	0.92	74	10	10.4	H
2.90	0.94	30	3	10.4	W
3.08	1.00	95	4	10.4	H
3.09	1.00	46	4	10.4	H
3.21	1.04	39	1	10.4	H
3.26	1.06	59	4	10.4	H
3.36	1.09	52	4	10.4	H
3.39	1.10	51	4	10.4	H
3.44	1.12	66	7	10.4	H
3.53	1.15	45	4	10.4	H
3.53	1.15	58	4	0.84	H
3.66	1.19	81	4	0.8	H
3.68	1.19	65	7	10.4	H
4.00	1.30	31	3	10.4	W
5.48	1.78	6	10	0.4	W
5.50	1.79	93	4	6	H
11.8	3.83	75	3	0.8	H
12.8	4.16	5	10	0.4	W

Table 4. Average corrosion rates arranged in ascending order. Specimens of the AZ61 alloy in the as-received surface condition. (measurements: W. by weight loss; H, by hydrogen evolution)

Corrosion rate (mg/cm²/d)	Normalized corrosion rate	Specimen No	Days immersion	Exposed surface (cm²)	Measurement method
0.27	0.50	28	3	10.4	W
0.28	0.52	15	3	9	W
0.33	0.61	27	3	10.4	W
0.33	0.61	84	4	0.8	H
0.38	0.70	100	4	10.4	H
0.38	0.70	98	4	6	H
0.38	0.70	16	3	9	W
0.43	0.80	33	3	10.4	H
0.43	0.80	54	4	10.4	H
0.47	0.87	41	1	10.4	H
0.53	0.98	67	7	10.4	H
0.53	0.98	99	4	10.4	H
0.55	1.08	97	4	6	H
0.56	1.04	24	10	10.5	W
0.58	1.07	23	10	10.5	W
0.58	1.07	42	1	10.4	H
0.61	1.13	43	4	10.4	H
0.70	1.30	34	3	10.4	W
0.81	1.50	68	7	10.4	H
0.84	1.56	32	3	10.4	W
1.25	2.31	83	4	0.8	H
1.73	3.20	8	10	0.4	W
3.10	5.74	40	1	10.4	H
49.3	91.3	7	10	0.4	W

Table 5. Average corrosion rates arranged in ascending order. Specimens of the AZ61 alloy in the as-polished surface condition (measurements: W, by weight loss; H, by hydrogen evolution).

3.2. Variability in the results

The relatively moderate differences in the corrosion rate (factors of about 1.5-2) found in repeated tests are not surprising bearing in mind the random nature of the attack, as well as the possible uncontrolled small changes in the experimental conditions of tests. In general, the attack tends to be concentrated in areas where the protective action of the oxide film on the metallic surface is lowest due to lacks of continuity or lower stability. It is logical for the corrosion process to depend on the distribution and importance of defect sites in the oxide film and the distribution of passive/active states at points on the exposed surface. Observation of the corroded surfaces has revealed a lack of homogeneity in the attack, in which areas of uniform attack can coincide with areas of heterogeneous corrosion, isolated pitting and/or clusters of filiform corrosion [4].

3.3. Anomalous behavior

Apart from the above–mentioned variability of the results, which can be considered as "normal", the aspect that particularly draws attention is the high corrosion rate values-far removed from the mass of data-that sporadically and unpredictably appears in the corrosion tests. For the purposes of the present work, the term "anomalous" will be applied whenever the corrosion rates exceed at least triple the normal results defined by the median values, taking normality in the statistical sense.

3.4. Effect of surface state and alloy type

Through this analysis of the possible influence of diverse factors on the variability of the corrosion results, one of the factors considered has been the surface state of the tested specimens

In this work, corrosion rates for the AZ31 and AZ61 magnesium alloys immersed in the 0.6 M NaCl showed values mostly between 0.5-1 $mg/cm^2/d$ for AZ31 alloy and 0.5-3 $mg/cm^2/d$ for AZ61. These corrosion rates are summarized in Tables 2-5, where the data are set out in increasing order. The median values (middle value in the list of numbers) for the AZ31 alloy specimens is 0.64 $mg/cm^2/d$ in the as-received condition and 0.62 $mg/cm^2/d$ in polished surface condition. Much more noticeable is the effect of surface state for the AZ61 specimens, with median values of 3.08 $mg/cm^2/d$ for the as-received condition and 0.54 $mg/cm^2/d$ for the polished condition. The higher corrosion rate in the as-received condition for the AZ61 alloy suggests a combined effect of an oxide film of lower quality and an enhanced micro-galvanic action due to the abundant presence of β-phase in this alloy.

From the analysis of Tables 2-5 and Table 6 it can be deduced also that the AZ31 alloy specimens present a greater tendency towards anomalous behavior (7 cases in 45 tested specimens) than the AZ61 specimens (3 cases in 57 tested specimens), and that the frequency of cases of anomalous behavior does not depend exclusively on the alloy type, as it is notably lower for the AZ31 alloy in polished surface condition than in the as-received condition.

Testing variables	Occurrence of anomalous behavior
Alloy type:	
AZ31	7 cases per 45 tested specimens
AZ61	3 cases per 57 tested specimens
Surface condition:	
As-received surface	8 cases per 59 tested specimens
As-polished surface	2 cases per 42 tested specimens
Exposed surface area defined by the contour of a circular gasket	
0.5 cm^2	3 cases per 8 tested specimens
0.8 cm^2	6 cases per 14 tested specimens
9.0 cm^2	1 case per 8 tested specimens
Specimen of 2 cm x 2 cm x 0.3cm suspended vertically	
	5 cases per 64 tested specimens

Table 6. Frequency of cases of anomalous behavior for different exposure conditions and composition of the specimens.

3.5. Effect of exposed surface area

The effect of the geometric surface area was studied with specimens of the alloys AZ31 and AZ61 exposing a free surface areas of 0.5 cm^2, 0.8 cm^2 and 9.0 cm^2, delimited by the contour of circular plastic gaskets on the specimens in horizontal position. The results obtained, summarized in Table 6, shows a propensity to abnormal behaviour when the exposed area of the flat specimens is ≤ 0.8 cm^2. In this regard, Fig. 2 is an example of different morphologies of attack associated with the exposed surface area, being notably the presence of large irregular pits in the case of the small surface area exposed.

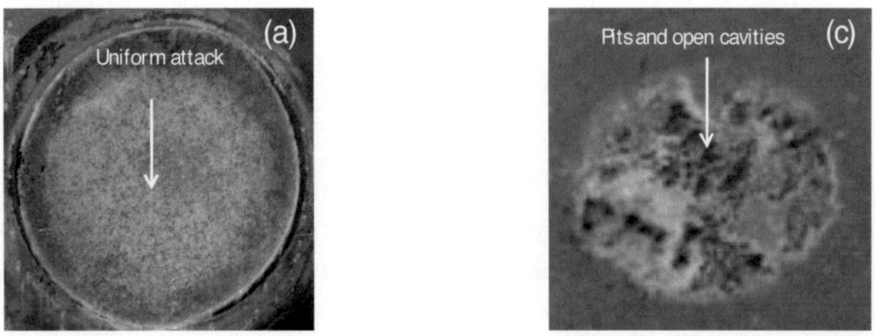

Figure 2. Surface appearance after immersion in 0.6M NaCl solution for 14 days at 24 °C and after removal of corrosion products. AZ31 specimen with the original surface in the as-received condition, Exposed circular area of ~9 cm^2 X 0.75 approxExposed circular area of ~0.4 cm^2 X 0.75 approx

Shi and Atrens [5] have reported that the corrosion rate of magnesium can be affected by the presence of crevices between the metallic specimen and the epoxy resin mounting in which is embedded, which can cause increases of the order of 50-100% in the corrosion rate. This suggests a possible crevice effect between the plastic gasket and the metallic surface. However, there are two reasons to doubt whether the anomalous behavior mentioned in this work can be explained by such a phenomenon: (i) the much greater magnitude of the corrosion rate that defines anomalous behaviour and (ii) the fact that notable attack of the metal is not observed under the plastic gasket.

Table 6 also suggests an almost insignificant tendency towards anomalous behaviour in the case of the specimens immersed in vertical position with their entire surface (~10.4 cm^2) exposed to the saline medium. One practical conclusion that may be drawn from Table 6 is that an area of around 10 cm^2 or greater reduces the risk of the appearance of anomalous results.

It is logical to relate an increase in the corrosion rate with the weakening, breaking up or disappearance of the oxide film that usually protects the surface of magnesium alloys. Any increase in the percentage of active points or oxide-free areas on the metallic surface will favour the tendency (stimulated by the presence of Cl^- ions) for metallic ions to pass from the reactive bare surface into the aqueous solution. Anomalous behaviour will probably be triggered in special circumstances when a large highly reactive area appears on the surface. Restoration of the passive oxide film would obviously inhibit anomalous corrosion, and so accelerated dissolution will only proceed if the surface is permanently activated by a process that reduces the thin oxide film which spontaneously tends to form on the metallic surface in contact with air.

A specific cause in the present work that has been favourable to the appearance of anomalous corrosion is the presence of a gasket that delimits a small exposed area upon which a persistent crowded population of hydrogen bubbles is likely to develop and maintain during the corrosion process, This suggests a possible association between circumstances related with the evolution of hydrogen and the activation of isolated regions on the metallic surface, which cease to be controlled by the oxide film mechanism. This film may be ruptured by the combined action of Cl^- ions and the extremely reactive nascent hydrogen, with the formation of a reducing atmosphere in which metal oxidation and restoration of the passive film is not possible. It may be speculated that factors related with the formation, permanence and disturbance of the hydrogen bubble layer exert some influence on the incubation of the above phenomenon.

The greater tendency for specimens with a small exposed area delimited by a circular gasket to behave anomalously in this work suggests the influence of this configuration on the special properties of the evolved hydrogen bubbles. Supposing a role of bubble removal, an effect of specimen geometry and hydrodynamics of the liquid environment may be expected.

The greater adhesion of bubbles on a rough surface is translated into a prolongation of the bubble residence time [6]. Perhaps connected with this fact is the more likely occurrence (Table 7) of anomalous behavior on the significantly rougher as-received surfaces.

http://dx.doi.org/10.5772/58370

Specimen			Corrosion rate data (mg/cm²/d)	
Nº	Alloy	Surface (*) condition	electrochemical	Gravimetric
1	AZ31	O	0.20	0.51
2	AZ31	O	0.25	1.84
3	AZ31	P	0.38	0.42
4	AZ31	P	0.40	0.50
5	AZ61	O	1.60	10.25
6	AZ61	O	2.85	4.38
7	AZ61	P	1.68	39.4
8	AZ61	P	1.64	1.38
9	AZ31	O	2.81	3.05
10	AZ31	O	0.81	0.84
11	AZ31	P	0.71	0.71
12	AZ31	P	0.75	0.71
13	AZ61	O	1.71	1.61
14	AZ61	O	0.84	0.88
15	AZ61	P	0.36	0.28
16	AZ61	P	0.30	0.38

(*) O=as-received; P=as-polished

Table 7. Comparison of corrosion losses estimated from electrochemical and gravimetric measurements for AZ31 and AZ61 alloys.

3.6. Underestimated electrochemical data

Different studies have shown that serious discrepancies may occur between electrochemical and gravimetric determinations of the corrosion rate of magnesium alloys, resulting in much lower electrochemical values than the real values determined gravimetrically [3, 7, 9]. In the literature, this behaviour has tentatively been associated with several phenomena, such as material disintegration, the participation of monovalent Mg^{+}ions in the corrosion process, and anomalous chemical dissolution, although the true cause of the large discrepancies remains to be fully clarified [10-14].

Logically the above leads to uncertainty with regard to the application of electrochemical methods to estimate the corrosion rate of magnesium alloys, especially in media containing chloride ions.

Using a series of 16 specimens (Nos. 1-16 in Table 7), electrochemical and gravimetric corrosion rate data has been simultaneously obtained with them to allow comparison between the two types of measurements. Fig.3 constructed with this data shows satisfactory consistency between both types of corrosion rate determinations for 12 of the 16 tested specimens, but not for the other four (specimens 1, 2, 5 and 7), where notable differences were seen; with electrochemical estimations of up to 20 times lower than the real corrosion rates determined by gravimetric measurements.

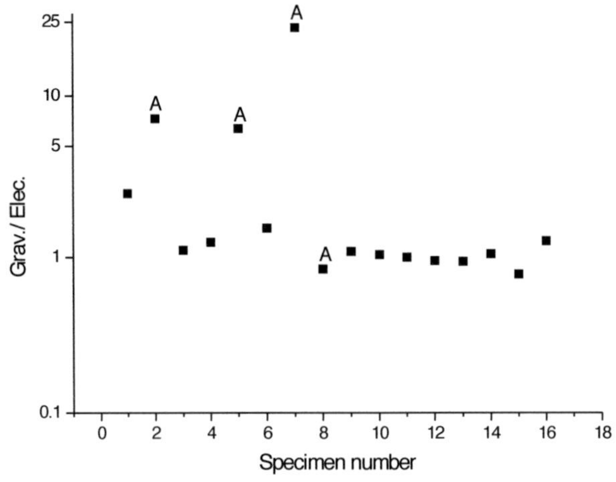

Figure 3. Ratio between gravimetric (GRAV.) and electrochemical (ELEC.) data. The points marked with the letter A correspond to specimens that showed anomalous dissolution rates in weight loss measurements.

An explanation for these cases of strong discrepancy is to accept that only a fraction of the metallic surface really contributes to the electrochemical response; for instance when the material simultaneously disintegrates into fine metallic particles during the corrosive attack [10-12] or when part of the corrosion process is chemical rather than electrochemical, so the simultaneous independent chemical dissolution processes cannot be monitored by electrochemical means [8, 9].

In the case of abundant hydrogen evolution, as in the attack of magnesium alloys in 0.6 M NaCl, the possibility has also been mentioned that gas bubbles adhering to the metal surface and insulating a fraction of it from the electrical signal applied during electrochemical measurements substantially affect the system´s response, as electrochemical data does not account for the entire corrosion process on the exposed surface [5, 15].

It is interesting to note that in three of the four specimens on which great differences have been seen between gravimetric and electrochemical weigh loss determinations (specimens 2, 5 and 7 in Fig.3), this behaviour has coincided with the occurrence of anomalous corrosion rates. The coincidence of both phenomena suggests that the electrochemical corrosion process only partly controls metallic dissolution in the case of anomalous behaviour, and that along with the normal corrosion reaction another non-electrochemical mechanism must play an essential role in the accelerated corrosion process.

http://dx.doi.org/10.5772/58370

3.7. Effect of pre-existing film and time of exposure to the test solution

This last section is concerned with the possible effect on the corrosion rate measurements, of (i) the native oxide films (just a few nanometers thick) which is spontaneously formed in the air and that may have some protective effect (barrier effect) against corrosion, and (ii) duration of exposure of the specimen under study to the test solution.

The effect of said oxide film seems to depend on the surface condition of metal. It has been found that this effect is more pronounced with the polished surfaces, chiefly of alloy AZ61. In the immersion tests in the 0.6 M NaCl solution a certain effect of the initial surface condition is revealed in the early stages of the test, principally with the polished specimens of the AZ61 alloy; in it polishing increased the corrosion resistance by a factor of about five whereas in the alloy AZ31 was only duplicated [16].

It is important to mention that, after an initial period which does not last for more than about 7 days, no effect attributable to the initial surface condition is generally seen. With the AZ61 alloy this effect has almost disappeared after 7 days. With the AZ31 alloy the effect of the air-formed film is smaller and is lost more slowly [16]. This different effect of surface condition are probably due to the fact that the oxide film formed on the polished surfaces is more uniform, homogeneous and continuous than that on the as-received surfaces resulting from the fabrication of the wrought alloy, on which the film will be discontinuous and porous and not uniformly distributed. From XPS determinations, it was found that the oxide film formed on de polished AZ61 alloy was some 2 nm thicker than on the as-received condition [16].

3.7.1. *Effect of time of exposure*

In order to facilitate the correct comparisons of corrodibility between specimens, it is desirable to take into account the effect of time of exposure to the corrosive solution; the assumption that corrosion occurs at a constant rate through all the test period may not in fact be true, since there are frequent cases in which corrosion rate tends to decrease or increase over time.

Often, rates of attack tend to diminish as a result of the formation of adhered insoluble corrosion products or protective films originated in the contact with environment. Several studies [17-20] report the formation of layers of corrosion products on the surface of the magnesium alloys immersed in corrosive solutions, which exert some influence on the kinetics of the corrosion process. For this reason, extrapolations of results of too short tests are more likely to indicate lesser resistance to corrosion than will actually be observed over a more prolonged period of exposure. The opposite case may also occur, for example as a result of the deterioration of protective oxide films on the specimen´s surface.

Fig.4, constructed with data from a previous investigation by Samaniego et al [4] may serve, as an example, to illustrate some of the features that can lead to erroneous ideas on the real values of corrosion resistance of the AZ31 and AZ61 magnesium alloys, Notable aspects include the marked reduction experienced by the instantaneous corrosion rate with immersion time for the AZ31 specimens and the tendency towards a slight decrease shown by the as-received surfaces of the AZ61 specimens after about four days immersion, effect explained by the presence of a slightly protective air-formed film on the as-received surface in conjunction with an accelerated micro-galvanic corrosion process.

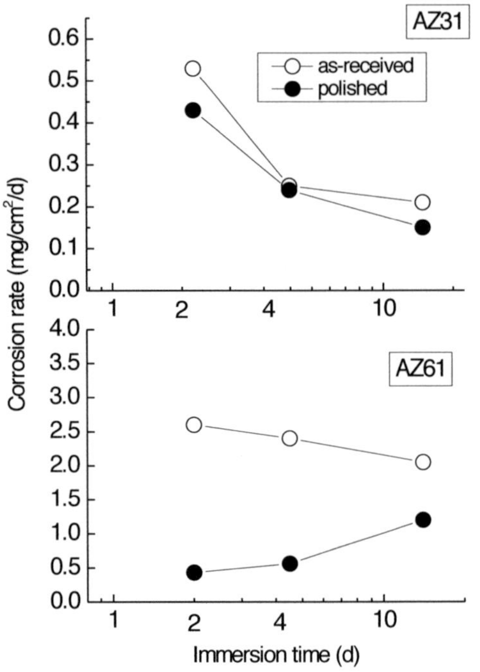

Figure 4. Instantaneous corrosion rate as a function of immersion time in 0.6M NaCl solution. Specimens exposing 10.4 cm^2.

4. Conclusions

1. In the immersion testing of alloys AZ31 and AZ61 in 0.6 M NaCl solution, anomalous results have occasionally been found in which the corrosion rate greatly surpasses normal values for the tested specimens.

2. A statistical type connection is seen between the frequency of the appearance of this anomalous result and certain characteristics of the tested specimens, in particular, the configuration and area exposed to the corrosive medium.

3. The results suggest some kind of link between anomalous behaviour and the action of hydrogen bubbles adhered to the specimen surface during the corrosion process.

http://dx.doi.org/10.5772/58370

4. The anomalous behaviour of the specimens tends to coincide with much smaller electrochemical corrosion rate determinations than those obtained by weight loss or hydrogen evolution measurements. This fact suggests that in the anomalous corrosion process only a fraction of the metallic surface really contributes to the electrochemical response.

5. On the basis of the information discussed, the conclusion is drawn that corrosion measurements may give rise to misleading comparisons of corrodibility without specifying the time during which the specimens to be compared are immersed in the corrosive solution.

Acknowledgements

We wish to express our gratitude to Prof. S. Feliu for several clarifying and stimulating discussions during the course of this work. Also, the authors gratefully acknowledge the financial support for this work from the Ministry of Science and Innovation of Spain (MAT 2009-13530)

Author details

Alejandro Samaniego and Sebastián Feliu (Jr)*

*Address all correspondence to: sfeliu@cenim.csic.es

Centro Nacional de Investigaciones Metalúrgicas CSIC, Madrid, Spain

References

[1] Song GL, Atrens A. Understanding Magnesium Corrosion-A Framework for Improved Alloy Performance. Advanced Engineering Materials 2003;5(12) 837-858.

[2] Kirkland NT, Birbilis N, Staiger MP. Assessing the Corrosion of Biodegradable Magnesium Implants: A Critical Review of Current Methodologies and their Limitations 2012; Acta Biomaterialia 2012;8(3) 925-936.

[3] Shi ZM, Lin M, Atrens A. Measurement of the Corrosion Rate of Magnesium Alloys Using Tafel Extrapolation. Corrosion Science 2010;52(2) 579-588.

[4] Samaniego A, Llorente I, Feliu Jr. S. Combined Effect of Composition and Surface Condition on Corrosion Behavior of Magnesium Alloys AZ31 and AZ61. Corrosion Science 2013;68 66-71.

[5] Shi ZM, Atrens A. An Innovative Specimen Configuration for the Study of Mg Corrosion. Corrosion Science 2011; 53(1) 226-246.

[6] Lumanauw D. Hydrogen Bubble Characterization in Alkaline Water Electrolysis. Mster Thesis. University of Toronto; 2000.

[7] Pardo A, Feliu Jr. S, Merino MC, Arrabal R, Matykina E. Electrochemical Estimation of the Corrosion Rate of Magnesium/Aluminium Alloys. International Journal of Corrosion 2010, http://www.hindawi.com/journals/ijc/2010/953850/.

[8] Drazic DM, Popic JP. Anomalous Dissolution of Metals and Chemical Corrosion. Journal of the Serbian Chemical Society 2005; 70(3) 489-511.

[9] Kolotyrkin YM, Florianovich GM. Anomalous Metal Dissolution: Experimental Facts and Theoretical Interpretation. Zashchita Metallov 1984; 20(1) 14-24.

[10] Makar GL, Kruger J. Corrosion Studies of Rapidly Solidified Magnesium Alloys. Journal of the Electrochemical Society 1990; 137(2) 414-421.

[11] Straumanis ME, Bhatia BK. Disintegration of Magnesium while Dissolving Anodically in Neutral and Acidic Solutions. Journal of the Electrochemical Society 1963; 110(5) 357-360.

[12] James WJ, Straumanis ME, Bhatia BK, Jhonson JW. The Difference Effect on Magnesium Dissolving in Acids. Journal of the Electrochemical Society 1963; 110(11) 1117-1120.

[13] Hoey GR, Cohen M. Corrosion of Anodically and Cathodically Polarized Magnesium in Aqueous Media. Journal of the Electrochemical Society 1958; 105(5) 245-250.

[14] Song GL, Atrens A. Corrosion Mechanisms of Magnesium Alloys. Advanced Engineering Materials 1999; 1(1) 11-33.

[15] Vogt H. The Incremental Ohmic Resistance Caused by Bubbles Adhering to an Electrode. Journal of Applied Electrochemistry 1983; 13(1) 87-88.

[16] Feliu Jr. S, Maffiotte C, Samaniego A, Galván JC, Barranco V. Effect of Naturally Formed Oxide Films and other Variables in the Early Stages of Mg-alloy Corrosion in NaCl Solution. Electrochimica Acta 2011; 56(12) 4554-4565.

[17] Baril G, Pebere N. The Corrosion of Pure Magnesium in Aerated and Deaerated Sodium sulphate solutions. Corrosion Science 2001; 43(3) 471-484.

[18] Pardo A, Merino MC, Coy AE, Viejo F, Arrabal R, Feliu Jr. S. Influence of Microstructure and Composition on the Corrosion Behavior of Mg/Al Alloys in Chloride Media. Electrochimica Acta 2008; 53 (27) 7890-7902.

[19] Santamaría M, Di Quarto F, Zanna S, Marcus P. Initial Surface Film on Magnesium Metal: A Characterization by X-ray Photoelectron Spectroscopy (XPS) and Photocurrent Spectroscopy (PCS). Electrochimica Acta 2007; 53(3) 1314-1324.

http://dx.doi.org/10.5772/58370

[20] Pebere N, Riera C, Dabosi F. Investigation of magnesium corrosion in aerated sodium-sulfate solution by electrochemical impedance spectroscopy. Electrochimica Acta 1990; 35(2) 555-561.

Chapter 8

Mechanical and Acoustic Properties of Magnesium Alloys Based (Nano) Composites Prepared by Powder Metallurgical Routs

Zuzanka Trojanová, Pavel Lukáč, Zoltán Száraz and Zdeněk Drozd

Additional information is available at the end of the chapter

http://dx.doi.org/10.5772/57454

1. Introduction

Magnesium materials are excellent candidates for structural applications where low weight plays an important role. However, Mg and its alloys possess low stiffness and strength compared with aluminum alloys. One way to enhance the strength of Mg is to reinforce it with stronger particles or fibers of the second phase, essentially forming composite microstructures. Composites reinforced with micro-sized particulates of various materials were very often used to enhance the elastic modulus and mechanical properties. Addition of micro-particles, such as SiC, Y_2O_3, MgO, Al_2O_3 particles and CNT (carbon nanotubes) into Mg and Mg alloys has been shown to improve yield strength, modulus, hardness, fatigue and wear resistance, as well as damping properties and thermal stability (Száraz et.al., 2007, Moll et.al., 2000, Ferkel & Mordike, 2001, Trojanová et al., 1997). By scaling the particle size down to the nanometer scale, it has been shown that novel material properties can be obtained (Thostenston et al., 2005).

2. Materials processing

2.1. Preparation of Mg alloys reinforced with SiC particles

The AZ91, WE54 and Mg-8Li magnesium alloys reinforced with SiC particles with were processed by a powder metallurgy method. Mixing of the matrix alloys powders with SiC microparticles (mp) was carried out first in an asymmetrically moved mixer with subsequent milling in a ball mill. The powder was capsulated in magnesium containers and extruded at

400 °C using a 400 t horizontal extrusion press. The composite samples were not thermally treated. Microstructure of the as prepared WE54/SiC composite in introduced in Fig. 1a (the micrograph was taken perpendicular to the extrusion direction). SiC mp are non-uniformly distributed in the matrix; they form in many cases small clusters. The size of sharp bounded more or less uniaxial particles was approximately 9 μm and the grain size in the matrix about 3-4 μm. Light micrographs and transmission electron micrographs showed no pores in the composite and the binding between SiC particles and the matrix was perfect. No defects were found in the vicinity of SiC particles and no chemical reaction at the interface matrix/SiC particles was observed. Microstructure of the Mg8Li/SiC composite exhibits a mixture of two phases (hexagonal close packed α phase and body centred cubic β phase). In Fig. 2 light α and darker β phase are visible together with SiC particles The X-ray analysis revealed the relation between both phases as $\alpha : \beta = 55 : 45$. As the grain size the mean value of 5±2 μm was taken from both phases. Resulting materials contained different volume fraction of particles: AZ91 (13 vol.%), WE53 (13 vol.%) and Mg8Li (7% vol.%).

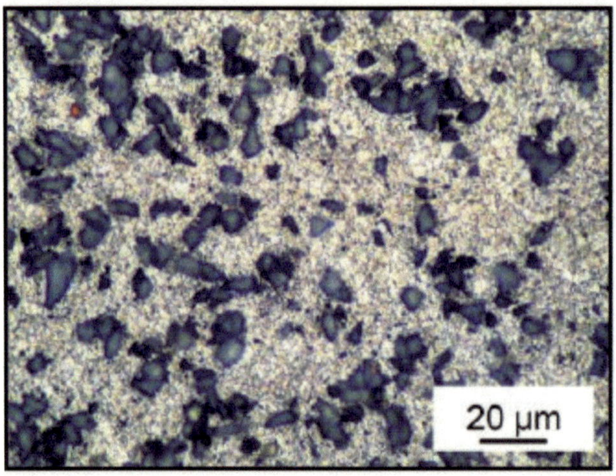

Figure 1. Microstructure of WE54/SiC.

2.2. Preparation of Mg based nanocomposites and ultrafine grained materials

Microcrystalline magnesium (μ-Mg) with 3 vol.% of alumina (Al_2O_3) nanoparticles (np) (μ-Mg +3nAl_2O_3), μ-Mg with 3 vol.% of zirconia np (μ-Mg+3ZrO_2) and microcrystalline Mg (μ-Mg) were studied. Micro-scaled Mg powder having a particle diameter of about 40 μm was prepared by gas atomisation of a high purity Mg melt in Ar atmosphere containing 1% oxygen for powder passivation. Both alumina and zirconia powders with a mean particle size of 14 nm were prepared by evaporation with the pulsed radiation of a 1000 W Nd:YAG laser and subsequent condensation of the laser-induced vapor in a controlled aggregation gas. The preparation method of np is described elsewhere (Naser et al., 1997, Ferkel & Mordike, 2001). The Mg powder was mixed with ceramic np in an asymmetrically moved mixer for 8 h. The

http://dx.doi.org/10.5772/57454

Figure 2. SEM micrograph of Mg8Li/SiC.

powder mixture was then milled together for 1 h in a planetary ball mill. Mixture was subsequently pre-compressed followed by hot extrusion at a temperature of 350 °C under a pressure of 150 MPa. After extrusion, the originally more or less equiaxial grains changed into elliptical grains with the long axis parallel to the extrusion direction. The grain size in the cross section was about 3 μm and in the extrusion direction 10 μm. The distribution of the np was not homogenous. The np or their agglomerates were located mainly along the grain boundaries of the resultant material. The transmission electron microscopical inspection revealed that only few particles were distributed within grains. Similar procedure was used for preparation of μ-Mg with 3 vol.% of alumina microparticles (μ-Mg+3μAl_2O_3)

Details of the preparation of the UFG-Mg and nanocomposite (nc) with the 3 vol.% of graphite nanopartiles (Gr np) (UFG-Mg+3nGr) was similar to the nc with the ceramic np. The micro-scaled Mg powder was prepared by gas atomisation of a high purity magnesium melt with argon atmosphere containing 1% oxygen for powder passivation. The Mg-powder had a median particle diameter of about 40 μm. The graphite powder used had a median particle size of 1-2 μm. The Mg powder was mixed with 3 vol.% of graphite powder in an asymmetrically moving mixer for 8 h. The powder mixtures were then milled for 8 h at 200 rpm in the planetary ball mill (Retsch, PM400) in a sealed argon atmosphere. The milling vessel of 500 ml volume was made of corundum and the milling balls (11 mm diameter) were made of hardened steel (100Cr6). The weight ratio of ball-to-powder was 10:1. The composite was encapsulated in an evacuated Mg container (70 mm in diameter), degassed at 350 °C, and extruded by the preheated (350 °C) 400 t horizontal extrusion press (outlet 14 mm). Analyses of the extruded material in an optical spark analyser (Spectrolab, Spectro Analytical Instruments) reveal no contamination of the composite by e.g. Fe from the extruder tools or milling balls, or Al from milling vessel (Ferkel, 2003). The mean grain size of specimens used was estimated, using transmission electron microscopy and X-ray profile analysis, to be about 150-200 nm. TEM of the UFG-Mg+3nGr is shown in Fig. 3.

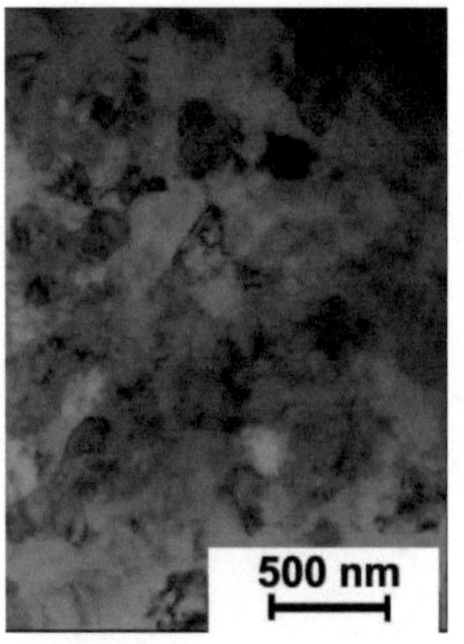

Figure 3. TEM of UFG-Mg with 3vol% Gr.

3. Mechanical properties

3.1. Composites reinforced with SiC microparticles

3.1.1. Compressive stress-strain curves

Deformation tests were carried out at temperatures between room temperature and 450 °C using an INSTRON testing machine. Cylindrical specimens of 8 mm diameter and 12 mm length were used for deformation tests performed in compression. The tensile specimens exhibited a gauge length of 10 mm and gauge diameter of 6 mm. Both samples were machined with a stress axis parallel to the extrusion direction. The strain rate sensitivity parameter *m* has been estimated by the abrupt strain rate changes (SRC) method. SRC tests and tensile tests with a constant strain rate ($\dot{\varepsilon}$= *const.*) were performed at temperatures from 300 to 450 °C. Temperature in the furnace was kept with an accuracy of ±1°C. Fig. 4 shows the compressive true stress-strain curves obtained for AZ91/SiC composite deformed at various temperatures. Samples were deformed either to fracture or at higher temperatures to predetermined strains. The stress-strain curves obtained at temperatures up to 150 °C exhibit the small hump at the beginning of deformation. Curves at temperatures higher than 150 °C are very flat; the maximum stress is achieved at lower strains. Similar stress strain curves obtained in compression for WE54+13%SiC composite at various temperatures are introduced in Fig. 5. The stress-strain curves obtained at higher temperatures have a flat character. The temperature influence on the strain hardening of the composite is well visible. The yield stress decreases with increasing temperature very slowly up to 200 °C. At temperatures higher than 200 °C both characteristic stresses (the compression yield stress and the compression strength) decrease substantially. It can be concluded that the thermal stability of the composite is up to 200 °C very good. The stress-strain curves estimated for Mg-8Li/SiC composite at various tempera-

http://dx.doi.org/10.5772/57454

tures are shown in Fig. 6. A continuous decrease in characteristic stresses can be seen from Fig. 6.

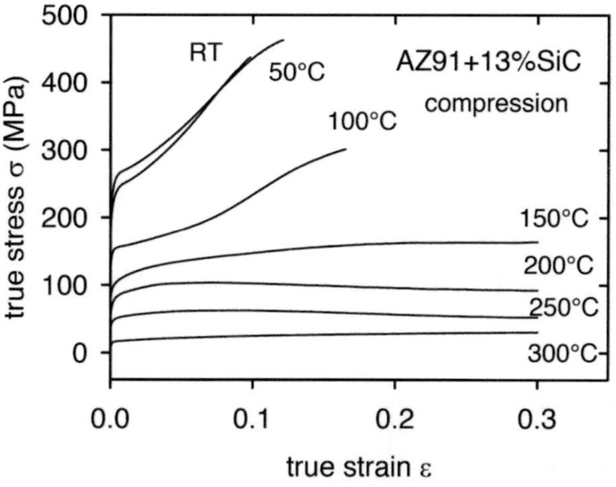

Figure 4. Compressive true stress-true-strain curves obtained for AZ91/SiC samples.

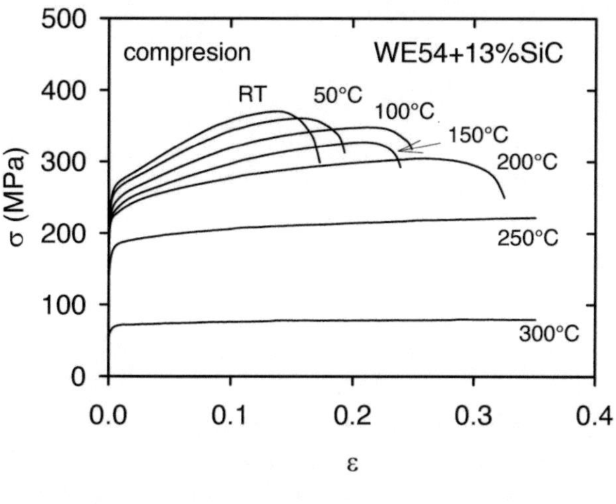

Figure 5. Compressive true stress-true strain curves obtained for WE54/SiC samples.

The compression yield stresses (CYS) estimated as the proof stresses at a strain of 0.002 are introduced in Table 1 together with the ultimate compression strength values (UCS). Three characteristic temperatures have been chosen.

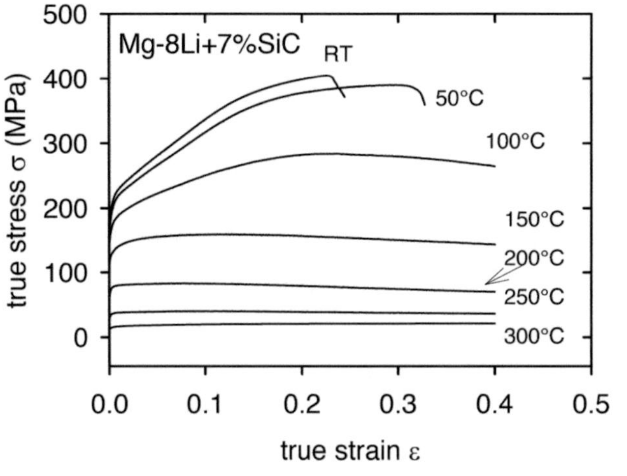

Figure 6. Compressive true stress-true strain curves obtained for Mg8Li.

	RT		150 °C		300 °C	
	CYS(MPa)	**UCS(MPa)**	**CYS(MPa)**	**UCS(MPa)**	**CYS(MPa)**	**UCS(MPa)**
AZ91/13SiC	222.5	437	98	163.8	15.6	30.6
WE54/13SiC	245.6	370	219.7	327.4	64.5	81.6
Mg8Li/7SiC	198.3	404.6	126.5	159	14.8	22

Table 1. Values of the yield stress and the ultimate compression strength estimated for three temperatures.

Table 1 shows a rapid decrease in the characteristic stresses obtained for AZ91/SiC and Mg8Li/SiC composites, while the decrease in the case of WE54/SiC composite is not so significant. This behaviour is very probably due to the presence of rare earth elements in the alloy. Rare earth elements form thermally stable precipitates situated in the grain boundaries. These grain boundary networks are also the reason for the good creep resistance.

TEM of the non-deformed WE54 composite sample and the WE54/SiC sample deformed in compression at 50 °C are shown in Figs. 7 and 8, respectively. TEM investigation showed that presence cuboidal particles which were identified as $Mg_{12}NdY$ precipitates. Many twins are a common feature for both non-deformed as well as deformed material. Grains in as-received material are well visible in Fig. 8. Thin twins, within single grains, are parallel to each other and quite narrow. Sometimes twins are extended through the grain boundary, causing a grain boundary deflection. The role of twinning is well known in deformation of *hcp* lattice alloys. Twinning is an important deformation mode of Mg alloys. It reorients the slip planes in order to relax stress concentrations and enhances multiple slip. Thin twins, often appearing in the form of parallel groups, where detected in compressed material (Ion et al., 1982). Another feature in the microstructure of the deformed composites is a high dislocation density.

http://dx.doi.org/10.5772/57454

Dislocations in many cases form pile ups and tangles (see Fig. 8). Significant amount of rectangular shape fine particles was estimated distributed within the grains.

Figure 7. Twins and precipitates in non-deformed composite. Dislocations in the right bottom corner are visible.

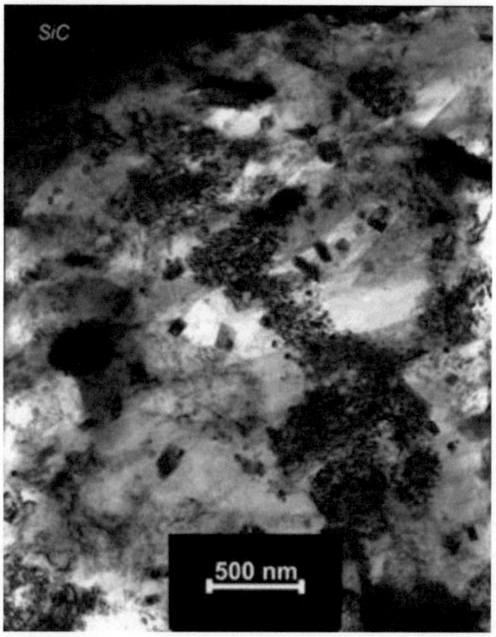

Figure 8. TEM of the WE54/SiC sample deformed at 50°C. SiC particle is situated in the top left corner of the picture.

In order to obtain more detailed information concerning the deformation mechanism(-s) occurring in particle reinforced magnesium alloys based composites, acoustic emission (AE) was used. AE stems from transient elastic waves which are generated within the material during deformation due to sudden localized and irreversible structure changes like dislocation glide and twinning, which may be considered as the main deformation mechanisms in Mg and its alloys due to their hexagonal crystal structure. WE54/SiC samples were deformed in compression at room temperature. The engineering stress- time plots together with the time variation of the AE count rates are shown in Fig. 9. Two AE maxima were observed. The observed AE maximum at the onset of plastic deformation is connected with the yield point (CYS ≅ 280 MPa) and it may be ascribed to the stochastic $\{10\bar{1}2\}\langle10\bar{1}0\rangle$ primary twin formation in the grains unfavourably oriented for the basal slip during the very early stage of plastic deformation. These twins reorient the original lattice on 86.3° and the subsequent straining may continue by the basal slip and secondary twinning.

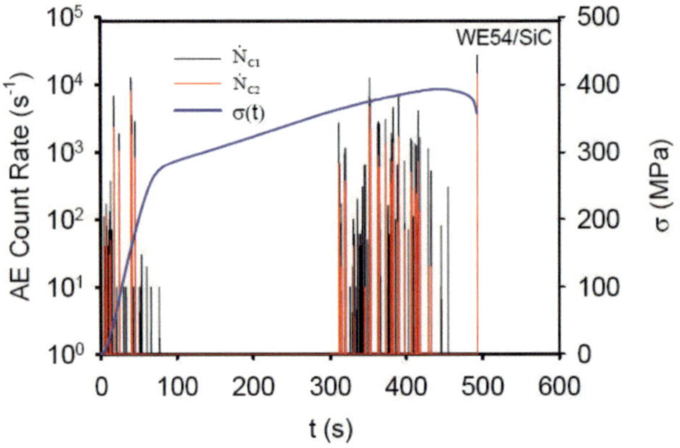

Figure 9. Deformation curve and AE count rate for WE54/SiC composite obtained at ambient temperature.

New twins were observed in the deformed microstructure (Fig. 10). Slim twins were observed also by the TEM as demonstrated in Fig. 8. These twins (and also dislocations visible in Fig. 8) arose very probably during extrusion of the powder material. Possible dislocation contribution to the AE signal is very probably marginal due to small grain sizes. The slip length of dislocations is very short (see dislocation tangles in the left bottom corner of the micrograph 8) and therefore the probability of pile ups formation is also very low. The mobility of dislocations is further limited due to the presence of small cuboidal particles visible in Fig. 7 and 8. The AE signal detected at the time point of ~310 s and onwards is discontinuous, and its sources may be ascribed to damage processes. Specifically, with the addition of the reinforcing phase, the geometrically necessary dislocations are generated to accommodate the plastic mismatch in the matrix. The stress concentrations in the vicinity of the reinforcing SiC particles may achieve their critical value and a breakage of particles and/or a release of cracks (decohesion) between the matrix and particles can occur. Both processes are considered as strong sources of the AE signal. The broken SiC particles and also cracks in the vicinity of particles were observed at the polished surface of the sample deformed to failure (see Fig. 11). Two main mechanisms were detected during plastic deformation: breakage of particles and decohesion in the particle–matrix interface. Based on these results we may consider localisation of the plastic deformation and fracture of the sample if the number of broken particles achieves its critical value.

Figure 10. Light micrographs showing new twins formed during plastic deformation in the WE54/SiC sample.

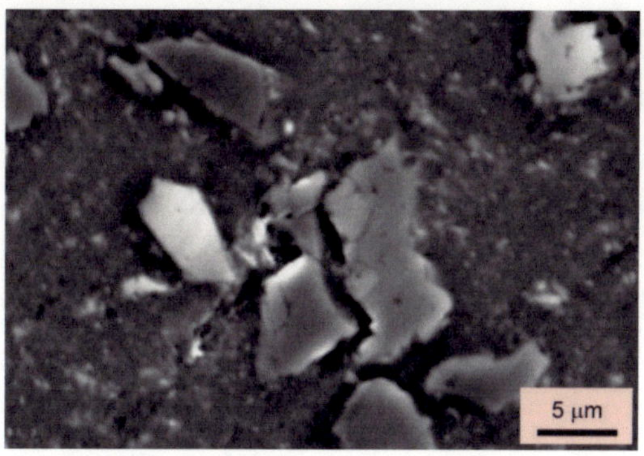

Figure 11. Scanning electron micrograph showing broken particles on the surface of the deformed WE54/SiC sample. Decohesion between the matrix and particles is also visible.

3.1.2. High temperature properties estimated in tension

The tensile true stress–strain curves estimated for WE54/SiC samples at a strain rate of 2.8×10^{-4} s^{-1} and various temperatures are shown in Fig. 12. A significant work hardening was obtained at temperatures up to 150 °C. A considerable decrease in the flow stress was observed after increasing the temperature from 250 to 300 °C. Curves obtained at temperatures higher than 250 °C have a flat character. It indicates new deformation process(-es) taking place at temperatures higher than 250 °C. Small grain sizes of powder metallurgically prepared materials indicate possibility of superplastic deformation. To check this eventuality, SRC tests were performed at temperatures from 350 to 450 °C. The values of the strain rate sensitivity *m* estimated for various strain rates and temperatures are given in Fig. 13. The strong strain rate dependence of the *m*-parameter is obvious from the picture. With increasing temperature the dependence is shifted to higher strain-rates. However, the *m* parameter slightly increases with temperature; the maximum values are in the vicinity of 0.3 as it is obvious from Table 2.

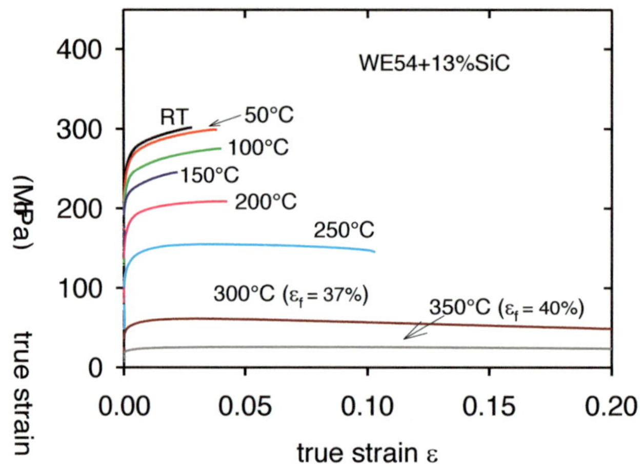

Figure 12. Tensile true stress-true strain curves estimated for WE54/SiC composite at various temperatures.

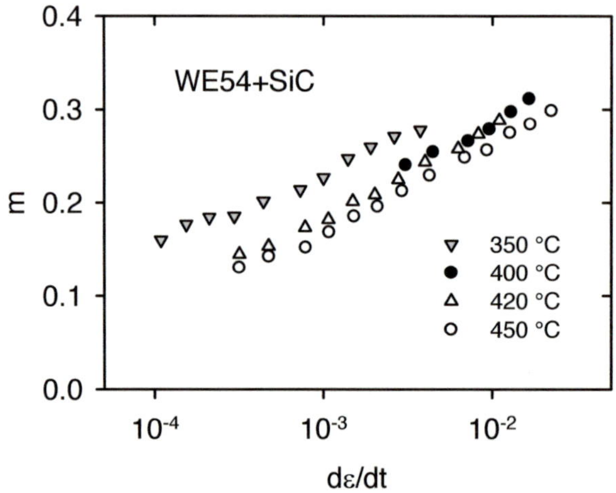

Figure 13. Strain rate dependence of m-parameter estimated at various temperatures.

http://dx.doi.org/10.5772/57454

The maximum recorded elongation to failure was 99%, which shows an evidence of the enhanced plasticity, nevertheless this value remains below the bottom limit for superplastic region. SEM micrograph of the sample exhibiting the highest elongation is shown on Fig. 14 documenting the microstructure after deformation at 450 °C. Numerous cavities formed during the high temperature deformation are visible in Fig. 14. The cavitation primarily occurred at the reinforcement/matrix interfaces, which are the preferential sites for the nucleation of cavities. The density of cavities was higher near the fracture surface where elongated cavities were found. Since many of these cavities were fairly large, it is reasonable to assume that growth and subsequent coalescence and interlinkage of the cavities led to the premature failure.

Figure 14. SEM micrograph showing cavities formed during high temperature deformation.

	m_{max}	$\dot{\varepsilon}$ (s^{-1})	ε_f (%)
350 °C	0.28	$4x10^{-3}$	66
420 °C	0.30	$2x10^{-2}$	95
450 °C	0.29	$5x10^{-2}$	99

Table 2. Maximum values of m-parameter and elongation to failure estimated for three temperatures and corresponding strain rates.

Similar study was performed for the Mg8Li/SiC samples. The strain rate sensitivity values *m*, obtained in tension using SRC method at temperatures 200, 250 and 300 °C are introduced in Fig. 15. It is obvious that the *m*–values increased with increasing deformation temperature and the dependence was shifted to the higher strain rates. While at 200 °C the maximum of *m* was reached at $\dot{\varepsilon}$= $4x10^{-5}$ s^{-1}, at a temperature of 300 °C the maximum laid in the vicinity of the strain rate $\dot{\varepsilon}$= $6x10^{-4}$ s^{-1}. The maximum value of *m* estimated at 300 °C exhibited 0.46, which is close to a value of 0.5, considered as an optimal value for the superplasticity. The maximum value

of $m = 0.3$ at 200 °C and all values of m estimated at temperatures higher for all strain rates were in the region enclosing the superplastic behaviour. The maximum ductility $A = 110\%$ was found at 300 °C and at a strain rate of $6x10^{-4}$ s^{-1}, which correspond to the maximum of the strain rate sensitivity $m = 0.46$. Despite of relatively high values of the strain rate sensitivity m, the achieved ductility was only at the onset of the superplastic region.

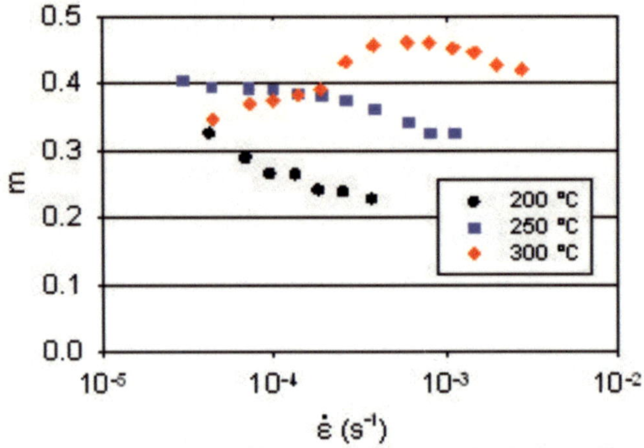

Figure 15. Strain rate dependence of the strain rate sensitivity parameter estimated for Mg8Li/SiC.

The observed cavities formation after high temperature deformation in WE54/SiC and Mg8Li/SiC composites indicates the presence of some diffusion process. The activation energy Q of such rate controlling is given by the relationship (Langdon, 1994):

$$\dot{\varepsilon} = \frac{AD_0Gb}{kT}\left(\frac{b}{d}\right)^p\left(\frac{\sigma}{G}\right)^n \exp\left(-\frac{Q}{RT}\right), \tag{1}$$

where $\dot{\varepsilon}$ is steady–state deformation rate, A is a dimensionless constant, d is the grain size, σ is the applied stress, p is the grain size exponent, n = 1/m is the stress exponent and D_0 is the pre–exponential factor; kT has its usual meaning and R is the gas constant. The activation energy Q is a slope of the plot σ/G vs $1/T$ (when $\dot{\varepsilon}T/G$ is constant):

$$Q = \frac{1}{m} R \frac{\Delta(\ln \sigma / G)}{\Delta(1/T)}. \tag{2}$$

The activation energy was estimated using relationship (2) for both composites to be Q(WE54/SiC) = 114kJ/mol. The activation energy for the volume diffusion is 135 kJ/mol and for the grain boundary diffusion 92 kJ/mol (Frost & Ashby, 1982). Comparing to an experimental value of 114 kJ/mol (using rule of mixture), we may conclude that the measured activation energy

http://dx.doi.org/10.5772/57454

consists of approximately 50% volume and 50% grain boundary diffusion. Using estimated values of m, the activation energy for Mg8Li/SiC composite was found (for the high m region) $Q \approx 87$ kJ/mol. According to binary Mg-Li diagram (Nayeb-Hashemi et al., 1984), the Mg–8Li alloy at 300 °C consists of equilibrium α phase and β phase whose chemical composition are 5.7 Li and 11 Li, respectively. Owing to higher mobility of Li in β phase the Li content in β phase decreases while in the α phase increases and the volume fraction of the α phase increases at the expense of β phase. Due to grain boundary migration caused by atomic mobility, the grain size of α phase increases.

Experimentally estimated value of the activation energy $Q = 87$ kJ/mol indicates that the main rate controlling mechanism is the grain boundary sliding accommodated with the grain boundary diffusion with the small contribution of the lattice diffusion. The successive grain growth increases the grain size in the α phase. The diffusion accommodation of the grain boundary sliding is more difficult, which implies cavities formation.

Observed formation and growth of cavities relaxes the stress concentration caused at the particles on the sliding grain boundaries. Cavities, created by vacancy clustering, may nucleate if the stress concentration is not relieved sufficiently rapidly. Local tensile stress caused by sliding at interfaces may be written in the form (Mabuchi & Higashi, 1999):

$$\sigma_{slid} = \frac{0.92\, kT\; d_p \dot{\varepsilon}\; d\; V_f}{\Omega\; D_L \left(1 + 5\dfrac{\delta D_{GB}}{d_p D_L}\right)}, \tag{3}$$

where d_p is the particle diameter, $\dot{\varepsilon}$ is the strain rate, d is the grain size, D_L is the lattice diffusion and D_{GB} is the grain boundary diffusion coefficient, δ is the grain boundary width, Ω is the atomic volume. V_f is the volume fraction of particles and kT has its usual meaning. The insufficiently accommodated grain boundary sliding process is the reason for cavitation and early failure of samples.

Based on our results of mechanical tests, microstructural observations and the analysis of the AE signal occurring during plastic deformation, we may conclude that the deformation processes in the powder metallurgically prepared magnesium alloys based composites are different at lower and higher temperatures. At lower temperatures (below about 150 °C), the deformation processes have following main characteristics:

i. Small grain size – several micrometers - is a typical feature of the microstructure.

ii. Early stages of the compressive plastic deformation are realised by twinning accommodated with the dislocation glide;

iii. Twin boundaries are impenetrable obstacles for the dislocation motion and contribute to a significant hardening at lower temperatures;

iv. Breakage of particles and decohesion in the particle–matrix interfaces cause localisation of plastic deformation and failure of materials;

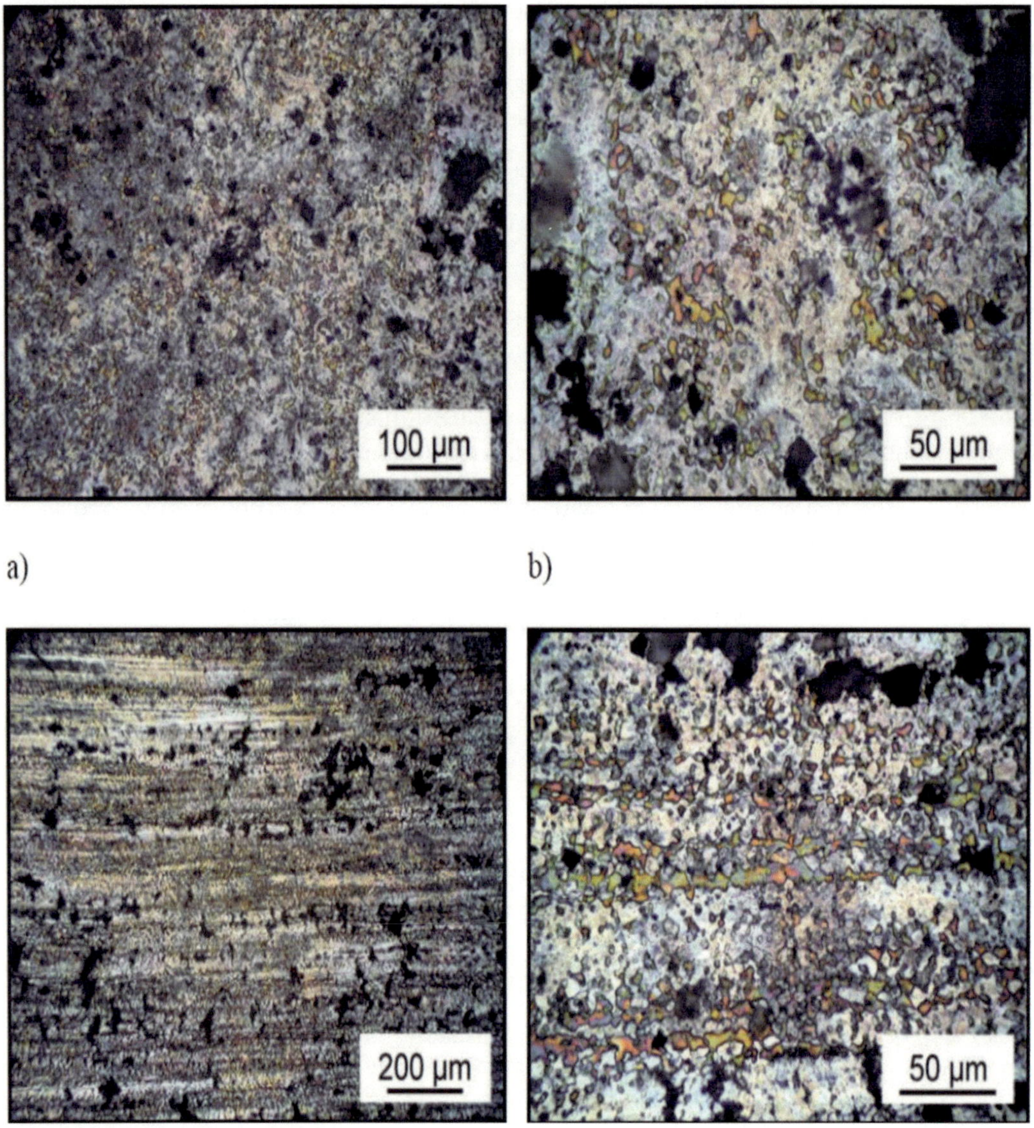

Figure 16. Light micrograph of Mg8Li/SiC composite after deformation at 300 °C and strain rate $6x10^{-4}$ s^{-1}, a), b) cross section, c), d) longitudinal section.

Although the grain size of composites exhibiting 3-5 μm indicated a possibility of the occurrence of the structural superplasticity, only enhanced plasticity was estimated at higher temperatures. This fact has following reasons:

i. The values of the strain rate sensitivity parameter reached m = 0.3-0.5;

ii. The grain boundary sliding is the significant mechanism operating at higher temperatures;

http://dx.doi.org/10.5772/57454

iii. The stress concentrations formed at the particles on the sliding grain boundaries are the reason for the cavity formation.

iv. Growth and subsequent coalescence and interlinkage of the cavities led to the premature failure of composites.

3.2. Stress-strain curves of magnesium micro- and nanocomposites

3.2.1. Microcrystalline magnesium reinforced with ceramic microparticles and nanoparticles

True stress-true strain curves obtained for microcrystalline Mg reinforced with alumina microparticles are introduced in Fig. 17a for tension and 17b for compression. A substantial difference between shapes of curves measured in tension and compression are visible. Similar curves were obtained also for microcrystalline Mg reinforced with alumina and zirconia np as it follows from Figs 18a,b and 19 a,b. It can be seen that the flow stress decreases as the test temperature is increasing. Substantial differences between curves obtained in tension and compression are obvious comparing Fig. 18a,b and 19a,b. While the curves measured in tension are mostly flat (for nc with alumina np at a temperature of 100 °C and higher), the curve estimated in compression exhibit local maxima: for μ-Mg+3Al_2O_3 at 100 °C and for μ-Mg+3ZrO_2 at room temperature and 100 °C. The main deformation mode in magnesium is the basal slip – glide of <a> dislocations. The secondary conservative slip may be realised by the motion of <a>–dislocations on prismatic and pyramidal planes of the first order. The basal <a> dislocations may react with the pyramidal <c+a> dislocations. Different dislocation reactions may produce both sessile and glissile dislocations (Trojanová et al., 2011). Production of sessile dislocations increases the density of obstacles for moving dislocations. Cross slip of <a> dislocations on prismatic planes and climb of <c> dislocations are the main dynamic recovery mechanisms. The flat course of the stress-strain curves is a result of a dynamic balance between hardening and softening. To fulfill the von Mises criterion for compatible deformation of polycrystals, the non-basal <c+a> slip and /or deformation twinning are needed (Yoo et al., 2001). In coarse grained Mg and Mg alloys, twinning is observed in the early stages of plastic deformation.

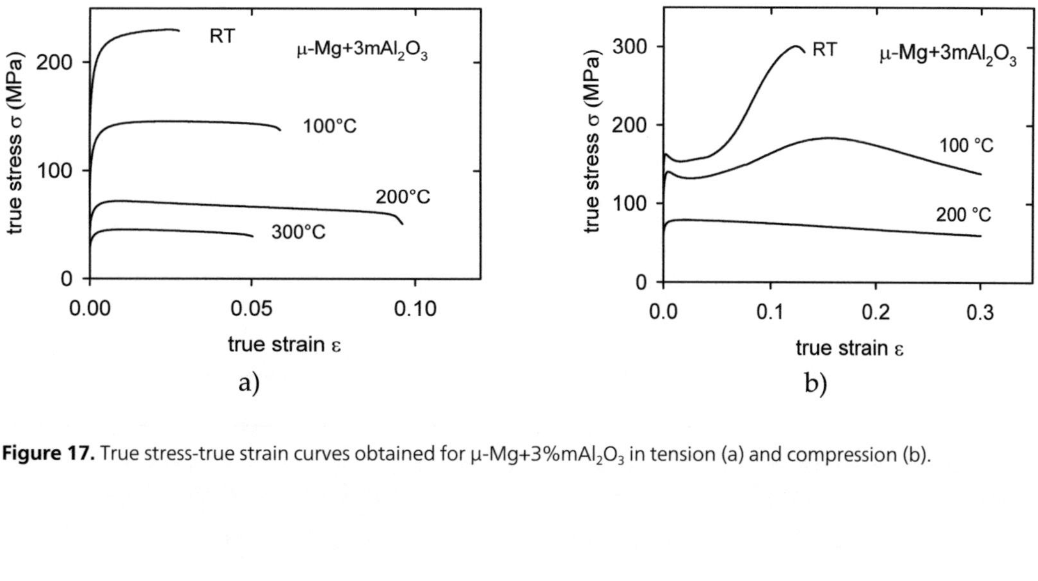

Figure 17. True stress-true strain curves obtained for μ-Mg+3%mAl_2O_3 in tension (a) and compression (b).

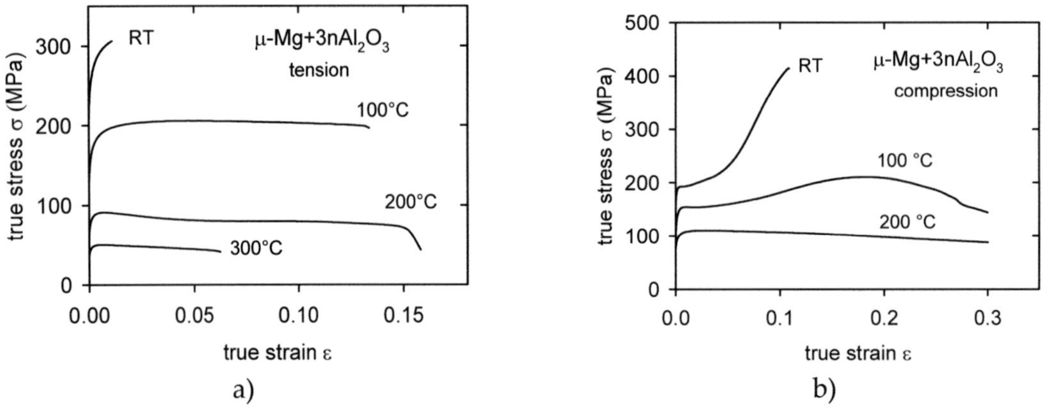

Figure 18. True stress-true strain curves obtained for μ-Mg+3%nAl_2O_3 in tension (a) and compression (b).

The stress-strain curves obtained in tension are flat at temperatures above 100 °C, there is a dynamic balance between hardening and softening. The work hardening rate is very close to zero. Tensile/compression yield stress asymmetry was studied by Mathis et al. using acoustic emission technique (Máthis et al., 2011). A broad maximum on the curves estimated at room temperature and at 100 °C was observed at a strain of about 10 %. Analysis of the acoustic emission signal revealed that during the compression tests below 200 °C, the maximum of the twin formation is at a very beginning of the stress-strain curve. The twin boundaries are obstacles for dislocation motion. Therefore, twin growth caused an increase in the flow stress, which was manifested by a rapid decrease in the acoustic emission signal. In tension, a significant twin formation was observed during the entire test. Twinning mode of deformation requires formation of the new interface inside of grains. The energy of twin interfaces in Mg is relatively high (Koike et al., 2003). This leads to an increasing difficulty of twin nucleation with decreasing grain size. This has been confirmed by the acoustic emission measurements (Trojanová et al., 2008).

T	**TYS (MPa)**				**UTS (MPa)**			
°C	**μ-Mg**	**μMg +3mAl_2O_3**	**μMg +3nAl_2O_3**	**μMg +3nZrO_2**	**μ-Mg**	**μMg +3mAl_2O_3**	**μMg +3nAl_2O_3**	**μMg +3nZrO_2**
20	229.0	200.0	269.0	175.5	256.5	230.0	306.2	207.5
100	145.7	128.2	172.6	114.8	161.2	145.4	205.5	125.8
200	77.6	65.9	86.0	68.4	84.5	71.5	91.1	73.2
300	48.5	41.5	48.5	46.2		45.3	50.3	51.0

Table 3. The tensile yield stress and ultimate tensile strength estimated at various temperatures in tension.

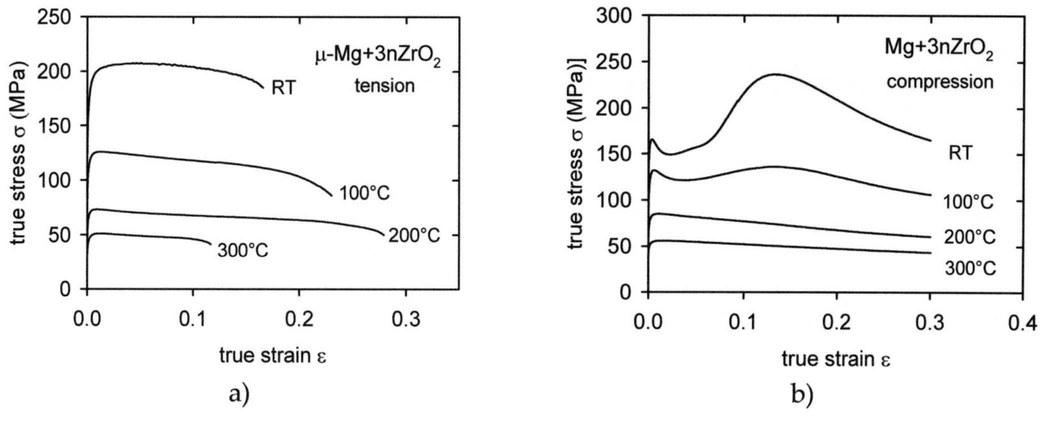

Figure 19. True stress-true strain curves estimated at various temperatures for.μ-Mg+3nZrO_2 sample in tension (a) and compression (b).

The temperature variations of the TYS and the UTS estimated for μ-Mg and μ-Mg with ceramic mp and np are introduced in Table 3. Relatively high values of the TYS and UTS were estimated. This is also influenced by small grain size of all materials as a result of the powder metallurgical preparation route. The Influence of the particles on the yield stress depends on the size of particles and kind of the np and, of course, on the test temperature. There are large differences in the values of the TYS at room temperature and 100 ºC. The difference in the values of the TYS for different composites measured at room temperature may be explained by the presence of the np and by the bonding between the particles and the matrix. The difference in the values of the TYS measured at 200 ºC is small and at 300 ºC, the values of the yield stress are the same for μ-Mg and μ-Mg reinforced with ZrO_2 and Al_2O_3 particles. Relatively high difference between the TYS and the UTS of μ-Mg+3Al_2O_3 and μ-Mg+3ZrO_2 (approximately 100 MPa) may be caused by the strength of the bonding between Mg matrix and ceramic np. While this bonding between Mg and Al_2O_3 np is nearly perfect, in the case of ZrO_2 is weak as it will be demonstrated in the paragraph 4.1.

T	CYS (MPa)			UCS (MPa)		
°C	μMg +3mAl_2O_3	μMg +3nAl_2O_3	μMg +3nZrO_2	μMg +3mAl_2O_3	μMg +3nAl_2O_3	μMg +3nZrO_2
20	162.8	188.9	162,1	300.8	415.7	236.4
100	135.2	140.8	124.7	183.7	210.6	136.2
200	72.2	96.0	79.6	79.0	110.5	85.0
300			52.2			56.9

Table 4. The tensile yield stress and ultimate tensile strength estimated for various temperatures in compression.

The temperature variations of the CYS and the UCS are introduced in Table 4 for Mg reinforced with various types of particles. It can be seen that while the values of the CYS of all materials studied are lower, the UCS values are higher than those measured in tension. The twinning mode of deformation in the early stages of deformation is easier in the compressive straining. New interfaces formed inside the grains are impenetrable obstacles for dislocation motion and they lead to hardening which is manifested with the local maximum on the stress-strain curve. The height of this maximum estimated at RT and 100 °C agree with the UCS value. The role of twinning decreases with increasing deformation temperature.

3.2.2. Ultrafine grained magnesium

Figure 20a shows the true stress-true strain curves for UFG-Mg deformed in compression at various temperatures. Samples were deformed either to fracture or at higher temperatures to predetermined strains. A pronounced upper yield point followed by the hardening stage was observed at temperatures between room temperature and 100 °C. The stress-strain curves at temperatures higher than 150 ºC have a steady state character. Stress-strain curves measured for UFG-Mg with 3 vol. % of Gr np (UFG-Mg+3nGr) in compression are presented in Fig. 20b for various temperatures. The local maximum at the beginning of the stress strain curve was in this case observed only at ambient temperature. The temperature dependences of the CYS and UCS estimated for both materials are shown in Table 5. Both stresses decrease rapidly with increasing temperature. The yielding phenomenon is very probably caused by the avalanche release of dislocations pinned at grain boundaries. It will be shown later in the paragraph 4.2 that the amplitude dependence of the logarithmic decrement in UFG-Mg indicates this phenomenon. Acoustic emission measurements detected no signal from the deformed sample at room temperature. This indicates that no dislocation pile-ups have been formed. In order to produce a pile-up of dislocations the grains need to contain a Frank-Read source operating in a cyclic way.

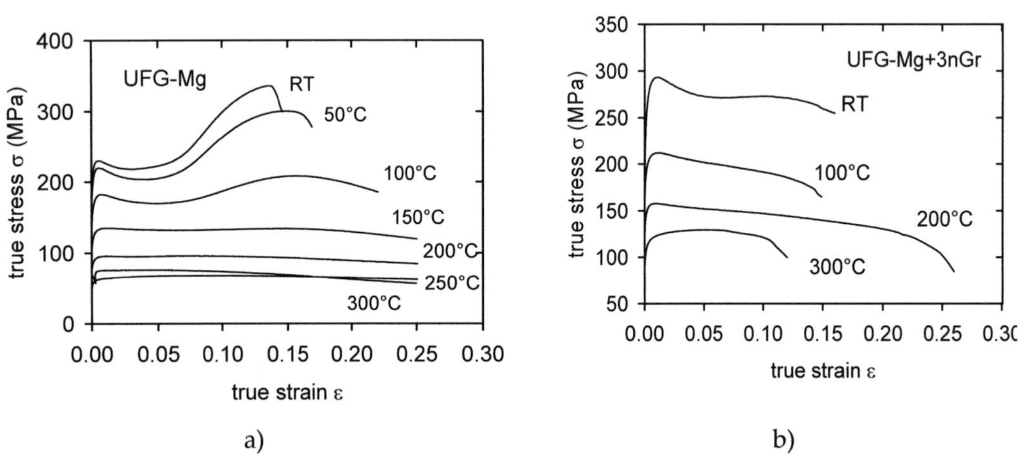

Figure 20. True stress-true strain curves obtained for various temperatures in compression for UFG-Mg (a) and UFG-Mg+3nGr (b).

Temperature	UFG-Mg compression		UFG-Mg+3Gr compression	
T (°C)	CYS (MPa)	UCS (MPa)	CYS (MPa)	UCS (MPa)
22	222.0	335.9	259.4	293.2
50	209.3	300.3		
77			196.3	212.0
100	170.2	208.3		
150	124.0	135.1		
200	89.5	95.7	149.3	157.9
250	59.6	76.1		
300	59.6	68.0	111.0	129.4

Table 5. The compressive yield stress and ultimate compression strength estimated for various temperatures for UFG-Mg and UFG_Mg with Gr np.

The maximum link length of a source to operate in a cyclic way becomes about D/2 (D is the separation of grain boundaries). The minimum resolved shear stress on the grain boundary due to the first loop can be written as

$$\tau_{GB} = \frac{2Gb}{D} - \frac{Gb}{D} = \frac{Gb}{D}; \tag{4}$$

where G is the shear modulus and b the Burgers vector of dislocations. The first term is the operating stress of the source and the second is the back stress from the first loop that is forced up against the grain boundary. If τ_{GB} exceeds the critical stress τ_{c}, required to an activation of an accommodating system, the first loop will escape in the grain boundary before the second loop is ejected from the source, and no pile-up will be formed (Nes et al., 2005). The stress-strain curves at temperatures higher than 150 °C exhibit a steady state character indicating a dynamic balance between hardening and softening.

It is clear that the Gr np increase the CYS. It is interesting to note that the UCS value of UFG-Mg deformed at RT is higher than that of Mg+3nGr. Values estimated at 300 °C for both materials show that the softening in the Mg+3nGr is lower than that in UFG-Mg. In the case where materials were prepared by milling, the refinement of the originally large grains into the UFG region is connected with large plastic deformation. This large plastic deformation is partially recovered during consolidation of the powder. Such mc and UFG materials usually contain a high density of dislocations in a heavy-deformation micro-structure (Mohamed, 2003). Np strengthening and grain refinement of the matrix are the key strengthening mechanisms in the UFG-Mg+3nGr nc (Lukáč et al., 2006, Estrin & Vinogradov, 2013) A TEM micrograph of the UFG-Mg+3nGr sample deformed at RT is shown in Fig. 21. Dislocations pinned at grain boundaries are visible. The influence of grain size on the flow stress of magnesium alloys has been subject of a number of investigations (e.g. Mabuchi et al., 2001,

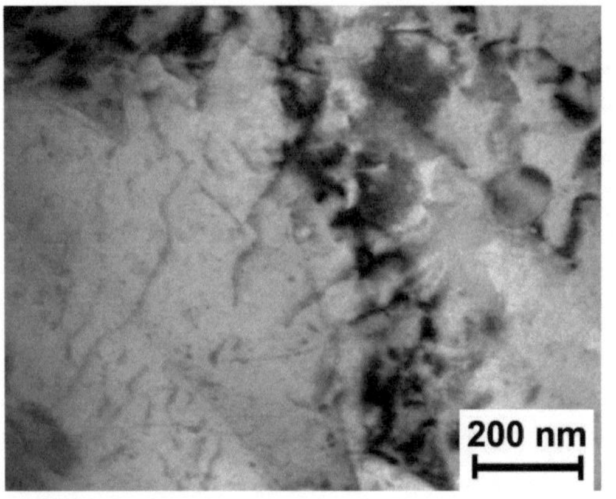

Figure 21. TEM of UFG-Mg+3nGr deformed at room temperature.

Anderson et al., 2003). Choi and co-workers (Choi et al., 2010) studied the deformation behaviour of ball milled Mg with the grain size from 60 nm up to 1 μm. They estimated that the YS is result of many mechanisms including the Hall-Petch strengthening, deformation twins, emission of partial dislocations and grain boundary sliding. Based on these estimations the deformation modes in the extruded Mg can be at RT divided into three regions:

(1) Mc region (the grain size above 1 μm): the yield stress of Mg follows the Hall-Petch relationship.

(2) UFG region (the grain size between 100 nm and 1 μm): the YS of Mg negatively deviates from the Hall-Petch relationship.

(3) Nc region (the grain size smaller than 100 nm): the slope of the yield stress dependence on the grain size is negative.

While for mc Mg the interaction of dislocations with grain boundaries and/or twin boundaries can be considered as significantly affecting plastic deformation, for UFG-Mg the dislocation pile-up mechanism loses its significance; twining and grain boundary sliding gradually contributes to plastic deformation.

3.3.3. X-ray analysis

Deformed UFG-Mg samples were investigated by the high resolution X-ray diffraction peak profile analysis. The diffraction profiles are evaluated by assuming that peak broadening is caused by small crystallites and strain caused by dislocations. The Williamson–Hall plots (integral breadth β vs. sinϑ, where ϑ is the Bragg angle) were constructed. The measured physical profiles are fitted by theoretical profiles calculated on the basis of well-established profile functions of size and strain. Both strain and strain anisotropy are accounted for by the dislocation model of lattice distortions (Wilkens, 1970; Ungár & Borbely, 1996). In a crystal containing dislocations, the mean square strain is done by Wilkens (1970)

$$\left\langle \varepsilon_{g,L}^{2} \right\rangle \cong \left(\rho C b^{2} / 4\pi \right) f(\eta), \tag{5}$$

where ϱ is the density of dislocations and b is their Burgers vector, C is the dislocation contrast factor, f(η) is the Wilkens function, where $\eta = L/R_e$ (R_e is the effective outer cut-off radius of dislocations), L is the Fourier length defined as $L = na_3$ ($a_3 = \lambda/2(\sin\vartheta_2 - \sin\vartheta_1)$), n are integers starting from zero, λ is the wavelength of X-rays and $(\vartheta_2 - \vartheta_1)$ is the angular range of the measured diffraction profile. In the case of hexagonal polycrystals, where dislocations with various Burgers vectors take place in deformation, the average contrast factor may be obtained by the weighted linear combination of the individual contrast factors for the active sub slip systems:

$$\left\langle Cb^{2} \right\rangle = \sum_{i=1}^{N} f_i \left\langle C^{i} \right\rangle b_i^{2} \tag{6}$$

where N is the number of the different activated sub slip systems, $\langle C^{i} \rangle$is the average dislocation contrast factor corresponding to the ith sub slip system and f_i are the fractions of the particular sub slip systems by which they contribute to the broadening of a specific reflection. The average contrast factors for a single sub slip system (hk.ℓ) in hexagonal crystals are (Dragomir & Ungár, 2002):

$$\left\langle C_{hk\ell} \right\rangle = \left\langle C_{hk.0} \right\rangle \left[1 + q_1 x + q_2 x^{2} \right] \tag{7}$$

where $x = (2/3)(\ell/ga)^2$, q_1 and q_2 are parameters depending on the elastic properties of the material, $C_{hk.0}$ is the average contrast factor corresponding to the hk.0 type reflections, *g* is the absolute value of the diffraction vector and *a* is the lattice constant in the basal plane. The q_1 and q_2 parameters and the values of $C_{hk.0}$ have been calculated for hexagonal crystals and compounds in (Dragomir & Ungár, 2002). Two Burgers vectors types were taken into account: $\langle a \rangle = \frac{1}{3}\langle \bar{2}110 \rangle$ and $\langle c + a \rangle = \frac{1}{3}\langle \bar{2}113 \rangle$. Using the scheme described by Kužel & Klimanek (1989), it is possible to take for the number N of sub-slip systems in eq. (12) with the <a> or <c+a> Burgers vector $N\langle a \rangle = 4$ and $N\langle c + a \rangle = 5$, respectively. Once the Burgers vector types are determined, the value of $\langle C_{hk.0} b^{2} \rangle$ and the dislocations density ϱ can also be calculated; for further details see (Kužel & Klimanek, 1989). The experimental values of q_1 and q_2 denoted as can be estimated by the whole profile fitting procedure (details see (Kužel, 1998)).

The Williamson–Hall plots in terms of the integral breadth β vs. the magnitude of the reciprocal lattice vector G (G represents sinϑ) are shown in Figs. 22 for samples deformed at various temperatures. Circles correspond to the values calculated from the experimental data (Kužel, 1998). Squares correspond to the values calculated on the basis of the model (Dragomir & Ungár, 2002). The fitting procedure allows to estimate the density of basal <a> dislocations, the

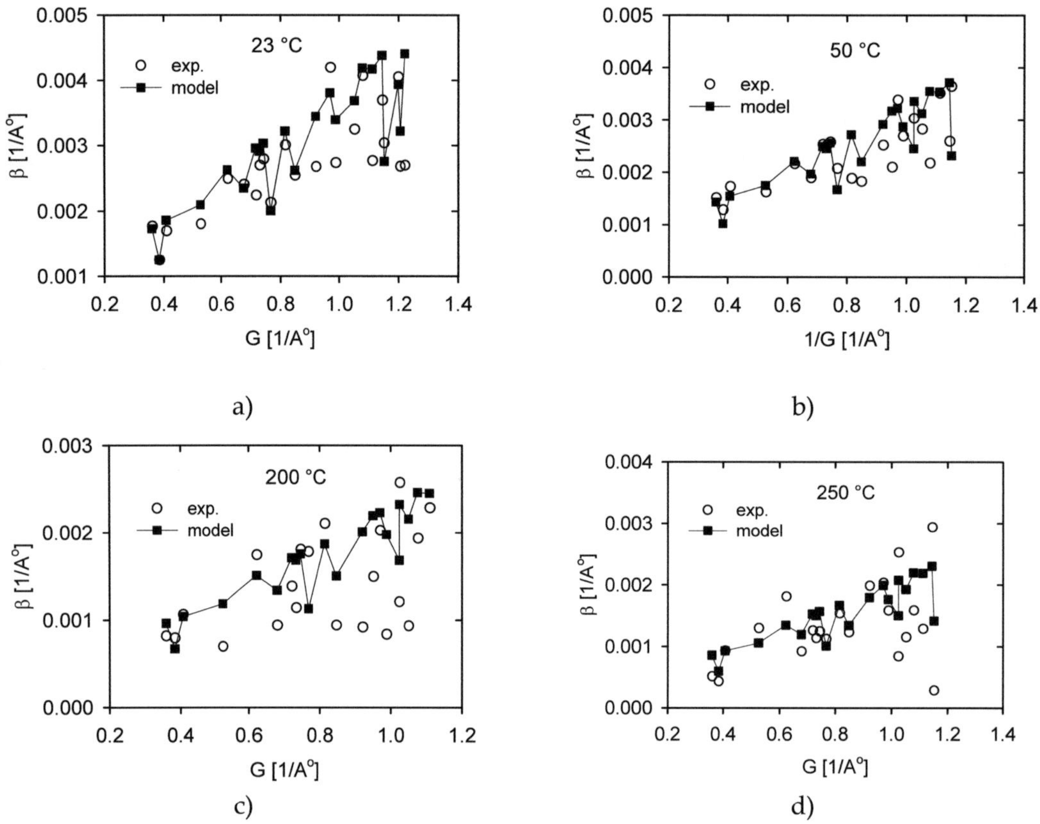

Figure 22. The integral breadths for the sample deformed at RT (a), 50 °C (b), 200 °C (c), 250 °C.

density of <c+a> dislocations and the grain size depending on the testing temperature. The results are given in Table 6. From Table 6 it follows that the basal dislocation density decreases with increasing deformation temperature, while the fraction of <c+a> dislocations is constant. The grain size growth has been also observed with increasing deformation temperature. Máthis et al., (2004) used the similar method to study the evolution of the dislocation density with the Burgers vector of different types as a function of the testing temperature in a coarse grained Mg. They found that at room temperature and at 100 °C, the dominant dislocations are <a> or mainly basal dislocations.

Deformation temperature (°C)	23	50	200	250
Density of basal dislocations $\rho_{<a>}$ 10^{14} (m^{-2})	4.04	3.0	1.5	1.2
$\rho_{<c+a>}/\rho_{<a>}$	0.1	0.1	0.1	0.1
Grain size (nm)	150	200	350	400

Table 6. Basal dislocation density $\rho_{<a>}$, density of <c+a> dislocations $\rho_{<c+a>}$ and grain size estimated for various deformation temperatures.

At higher temperatures, the fraction of <a> dislocations decreases, whereas the fraction of <c +a> dislocations increases. The fraction of <c> dislocations remains practically unchanged. The average total dislocation density increased considerably with strain as compared to the value in the as-cast state.

However, with increasing deformation temperature, the increment of the dislocation density decreased strongly in the plastically deformed sample. The concomitant increase of the fraction of <c+a> dislocations and the decrease of the average dislocation density with increasing deformation temperature can be explained by the dynamic recovery mechanism/-s. The results obtained in our study for UFG-Mg exhibit similar decrease in the basal dislocation density. The fraction of the <c+a> dislocations in UFG Mg remains constant, while it increases in the coarse grained material. This is very probably caused by the different deformation mechanisms in the UFG-material.

4. Internal friction measurements

4.1. Amplitude dependence of internal friction

The damping measurements were carried out in a vacuum (about 30 Pa) at room temperature. The specimens fixed at one end were excited into resonance by a permanent magnet fixed at the free side of the bending beam and a sinusoidal alternating magnetic field. The internal friction was characterised by the logarithmic decrement δ of the free decay of the vibrating beam $\delta = \ln (A_n/A_{n+1})$, where A_n and A_{n+1} are the amplitudes of a free decay of vibrations after n and (n+1) cycles, respectively. The resonance frequency ranged from 130 to 160 Hz. The strain amplitude dependences of the logarithmic decrement were measured. The specimens were annealed step by step at increasing temperatures up to 500 °C for 0.5 h and after each annealing cycle quenched into water of room temperature. The annealing at higher temperatures was performed in an argon atmosphere to avoid oxidation. The damping measurements were carried out immediately after heat treatment and quenching at room temperature.

Figure 23 shows the logarithmic decrement plotted against the logarithm of the maximum strain amplitude for μ-Mg+3nZrO_2 samples. The logarithmic decrement was measured before (as received - as rec) and after step by step annealing at increasing upper temperature of the cycle. It can be seen that the measured strain dependence of the logarithmic decrement exhibits two regions and can be expressed as

$$\delta(\varepsilon) = \delta_0 + \delta_H(\varepsilon), \tag{8}$$

where δ_0 is the amplitude independent component (or only weakly dependent on the maximum strain amplitude) found in the first region, for lower strain amplitudes. In the second region, for higher strain amplitudes, the component $\delta_H(\varepsilon)$ depends on the strain amplitude (amplitude dependent internal friction – ADIF); it increases with increasing strain amplitude. The upper temperature of the thermal cycling influences the logarithmic decrement. In the case of the μ-Mg+3nZrO_2 composite, the thermal treatment influences mainly the amplitude independent component of the decrement. It is important to note that very high values of damping in the amplitude independent component - in order of 10^{-2} - were obtained. The strain amplitude dependences of the logarithmic decrement for μ-Mg+3nAl_2O_3 are given in Figures 24. Figure 24a shows the strain amplitude dependence of the logarithmic decrement for specimens annealed up to 300 ºC. Figure 24b shows the strain amplitude dependence of the logarithmic decrement for specimens annealed at temperatures between 300 and 500 ºC. The thermal cycling influences mainly the amplitude dependent component δ_H that increases with increasing upper temperature of the cycle, if the upper temperature is between room temperature and 300 ºC. For the upper temperature between 300 and 500 ºC, the δ_H component decreases with increasing upper temperature.

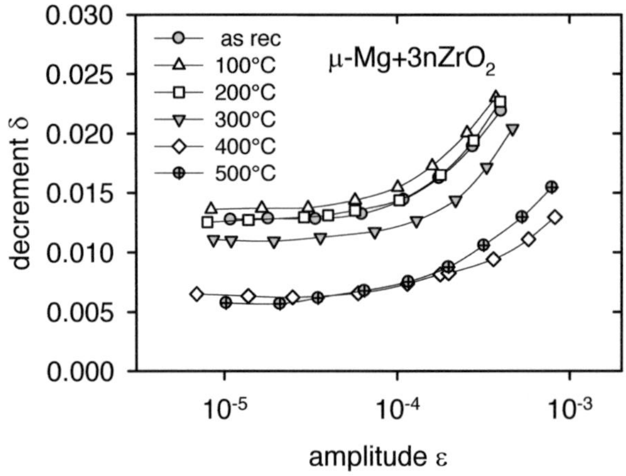

Figure 23. Amplitude dependence of decrement measured after annealing at increasing temperatures.

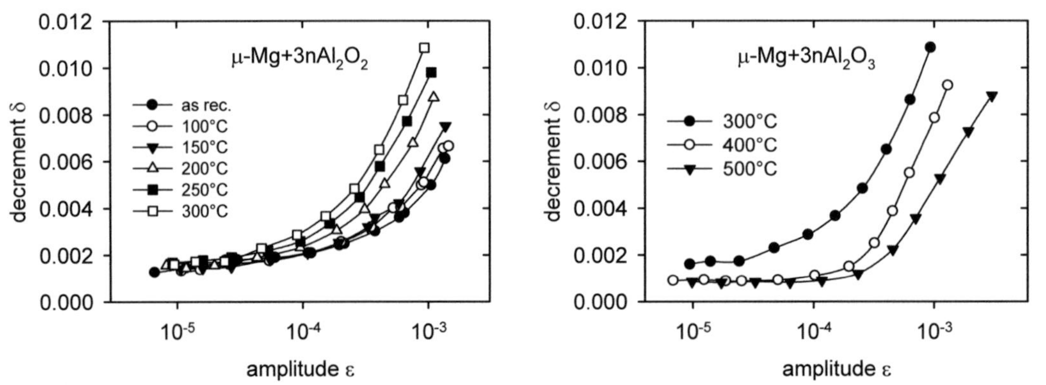

a) b)

Figure 24. Amplitude dependence of the decrement for μ-Mg+3nAl_2O_3 annealed at temperatures up to 300°C (a) and temperatures between 300 and 500 °C (b).

http://dx.doi.org/10.5772/57454

The strain amplitude dependences of the logarithmic decrement measured in UFG-Mg after thermal cycling are given in Figure 25a. Only a slight influence of thermal cycling on the δ versus ε curves can be observed. Figure 25b shows the plots of δ versus ε for UFG-Mg+3nGr. It can be seen that the thermal cycling causes a slight decrease of the logarithmic decrement at small strain amplitudes. The δ_H component changes only very slightly with increasing temperature of the cycle. Figure 26 shows the strain amplitude dependence of the logarithmic decrement for various materials before annealing (as received). The value of the critical strain ε_c at which the logarithmic decrement begins to increase with strain amplitude depends on annealing temperature and it depends on the kind of particles.

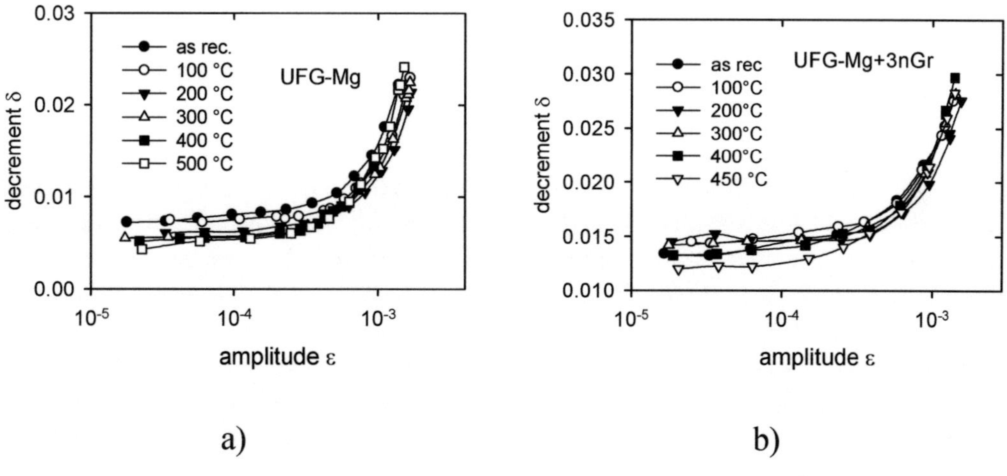

Figure 25. Strain amplitude dependence of the decrement for UFG-Mg (a) and UFG-Mg+3nGr (b), annealed at increasing temperature.

It is generally accepted that the δ_0 component is connected with the material microstructure and the amplitude dependent component δ_H is due to an interaction of dislocations with point defects (e.g. solute atoms). The dislocations contribute to the damping capacity by the motion of vibrating dislocation lines. Particulates (Al_2O_3, ZrO_2, Gr), as secondary phases, may improve the damping capacity of alloys. The particles in a composite may cause an increase in the dislocation density as a result of thermal strain mismatch between the ceramic particles and the matrix during preparation and/or thermal treatment. The difference between the coefficients of thermal expansion (CTE) of the particles and the matrix may create the thermal residual stresses after cooling from the processing temperature to room temperature. The thermal stresses are also generated during heat treatment and when composites are heated and cooled through a temperature range (thermal cycling). In generally, there are biaxial or triaxial stress fields located near the reinforcement-matrix interface. The amplitudes of the stress field decrease with increasing distance from the interface. In a simple one-dimensional model, the thermal stresses σ_{TS}, produced by a temperature change ΔT, at the interface are given by (Chawla, 1991)

$$\sigma_{TS} = \frac{E_F E_M}{\left(E_F f + E_M \left(1 - f\right)\right)} f\Delta\alpha\Delta T \tag{9}$$

where $\Delta\alpha$ is the mismatch of the CTE values between the reinforcements and the matrix, E_F and E_M are the values of the Young's modulus for the reinforcement and the matrix, respectively and f is the volume fraction of the reinforcement. The CTE of the matrix is much higher than that of the reinforcement. Thus after cooling, the tensile thermal residual stresses are created in the matrix. The thermal stresses may relax around the matrix-reinforcement interface by emitting dislocation. An increase in the dislocation density near reinforcement has been calculated as (Arsenault & Shi, 1986, Dunand & Mortensen, 1991):

$$\Delta\rho = \frac{B\, f\, \Delta\alpha\, \Delta T}{b(1-f)} \frac{1}{t} \tag{10}$$

where t is the minimum dimension of reinforcement, b is the magnitude of the Burgers vector of dislocations and B is a geometrical constant (depending on the aspect ratio). The newly formed dislocations may be sources of the higher damping capacity because of the motion of vibrating dislocation lines under cycling loading. According to the Granato-Lücke theory (Granato, & Lücke, 1981), weak and strong pinning points restrict motion of dislocations in their glide plane. At low strain amplitudes (stresses), dislocation bows out between the weak pinning points, leading to the amplitude independent logarithmic decrement. The dislocation remains anchored at the strong pinning points. At higher strain amplitudes, the force exerted on the weak pinning points becomes higher than the binding force of the weak pinning points. The dislocation segments break away from some weak pinning points within longer dislocation segments. This leads to a drastic instantaneous increase in the dislocation strain and thus giving rise to the level of the logarithmic decrement. The amplitude dependent component of the decrement depends on the dislocation density, length of shorter and longer dislocation segments. At temperatures higher than 0 K, the breakaway of shorter dislocation segments is thermally activated. During thermal cycling, new dislocations are created due to the difference in the CTEs (see Eq. (10)). An increase in the dislocation density, while the number of weak pinning points remains constant, increases effectively the length of dislocation segment between weak pinning points and hence the δ_H component should increase, which is observed. The higher temperature of the cycling, the higher the dislocation density and the higher increase in the effective length of the shorter dislocation segments. The critical strain ε_c at which the decrement becomes to be amplitude dependent is determined by the interaction energy between dislocations and the weak pinning points and the distance between the pinning points. As the distance between weak pinning point increases with the upper temperature, the dislocations can break away from the weak pinning point under lower critical stress and thus, the value of ε_c should decrease with increasing upper temperature, which is observed. The values of ε_c for a composite are proportional to the yield stress of the composite matrix. It should be consider that the value of ε_c can be influenced by the thermal stresses due to the difference in CTEs for both components of the composite. In mc and UFG Mg no new dislocations are created during thermal cycling. Hence, it is expected only a slight influence of the thermal cycling on the δ versus ε curves, which is in agreement with experimental results.

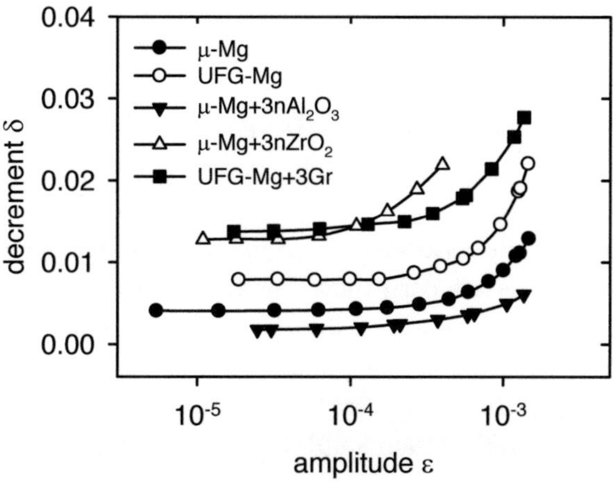

Figure 26. Amplitude dependence of the decrement for μ-Mg, UFG-Mg and μ-Mg and UFG-Mg reinforced with np.

Interfaces may contribute to the damping behaviour of composite materials. Due to a large difference in the coefficients of thermal expansion between the particles and the matrix, high thermal residual stresses are created around the particles. The thermal stresses with temperature can relax and plastic zones of tangled dislocation are formed. Their average radius depends on the difference in CTEs, particles radius and the change of temperature. The higher the matrix yield stress, the smaller radius of the plastic zones (Dunand & Mortensen, 1991). Plastic zones may also contribute to the damping behaviour.

The contribution of interfaces to internal friction depends on the character of interfacial bonding. The interface damping should be considered. The internal friction of a composite with the perfectly bonded interface (interface is incoherent) depends on the shape of particles, their volume fraction and the magnitude of local stress at the particle-matrix interface. Assuming that all the particles have the same diameter, the logarithmic decrement is given as (Zhang et al., 1994a)

$$\delta = \frac{4.5(1-\nu)}{\pi(2-\nu)} V_p \tag{11}$$

where V_p is the volume fraction of particles and ν is the Poisson ratio of the matrix. The effect of the perfectly bonded interfaces on the damping behaviour becomes more significant at high temperatures; in Mg based composites this could be above 150 ºC. The matrix becomes relatively soft relative to the nanoparticles. Hence, the internal friction at the interface may be thermally activated.

In the case of weak bonding at the interface, interfacial slip (sliding at the interface) may occur. In this case, the frictional energy loss caused by the sliding at interfaces may become a primary source of damping. The damping component due to interfacial slip under the applied stress amplitude σ_0 can be expressed as (Zhang et al., 1994a)

$$\delta = \frac{3\pi^2}{2} \frac{\kappa \sigma_r (\varepsilon_0 - \varepsilon_{cr})}{\sigma_0^2 / E_C} V_p \tag{12}$$

where κ is the friction coefficient between both components of the composite, σ_r is the radial stress at the particle-matrix interface corresponding to the stress amplitude σ_0, ε_0 is the strain amplitude corresponding to σ_0. The critical interface strain ε_{cr} corresponding to the critical interface shear stress τ_{cr} is the strain at which the sliding on the on interface begins. E_C is the elastic modulus of the composite. For weakly bonded interfaces, the critical strain ε_{cr} is assumed to be much lower than the strain amplitude ε_0 and then

$$\delta = \frac{3\pi^2}{2} \frac{\kappa \sigma_r}{\sigma_0} V_p. \tag{13}$$

The stress concentration factor σ_r/σ_0 has been reported to be 1.1-1.3 (Zhang et al., 1994b). The model does not take into account possible effects of temperature or frequency on damping and therefore the predictions of the model may be taken only as the first approximation. Very high values of δ_0 for mc and nc Mg reinforced by zirconia np and Gr np are very probably caused by interfacial slip due to weak bonding between particles and matrix. The effect of an additional damping due to the influence of particle can be clearly seen in Figure 26. Therefore, the effect of an additional damping due to the influence of particles may be estimated. The measured increase in δ_0 due to the addition of ZrO_2 np to μ-Mg and Gr np to UFG-Mg is about 0.008-0.009 (Figure 26). Taking for the friction coefficient a typical value of ~0.08, the relation (10) predicts a value for an additional damping owing to interfaces of $\delta_{0i} = 0.04$. The discrepancy between the predicted and measured value may be caused by a non-uniform strain state in specimens because they were subjected to bending in experiment. Hence, the stress (strain) level can reach its critical value at which the interface sliding starts only in some region of the measured specimen. Accordingly, the measured interface-damping component is lower than the predicted value. High values of the decrement component δ_0 obtained for mc and UFG-Mg are very probably caused by grain boundary sliding support by diffusion processes. On the other hand, the observed decrease in the amplitude independent component of μ-Mg +3nAl_2O_3 comparing with μ-Mg may be explained according to the following way: Np situated in grain boundaries prevent the grain boundary sliding which contribute significantly to the amplitude independent component of decrement.

http://dx.doi.org/10.5772/57454

4.2. Temperature relaxation spectrum of internal friction

The temperature dependent internal friction (TDIF), internal friction spectrum, is the temperature or frequency dependence of damping. Internal friction spectra were measured in a DMA 2980 (TA Instruments) apparatus in a single heating and cooling process. Measurements were performed at various frequencies from 0.1 up to 2 Hz. Throughout the measurements the strain amplitude was $1.2x10^{-4}$. The heating as well as cooling rate was 1 K/min.

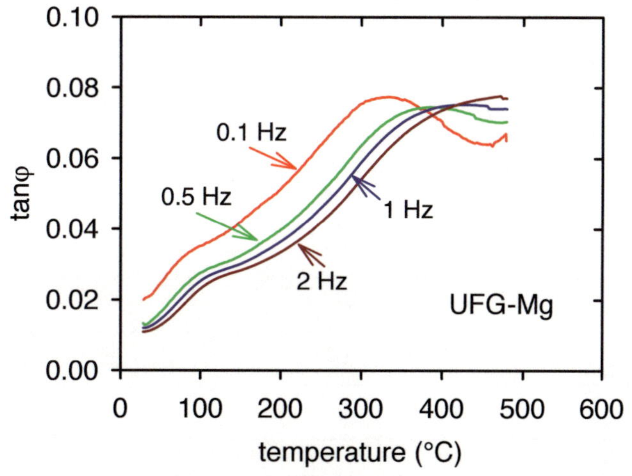

Figure 27. Temperature relaxation spectrum of IF estimated for various frequencies.

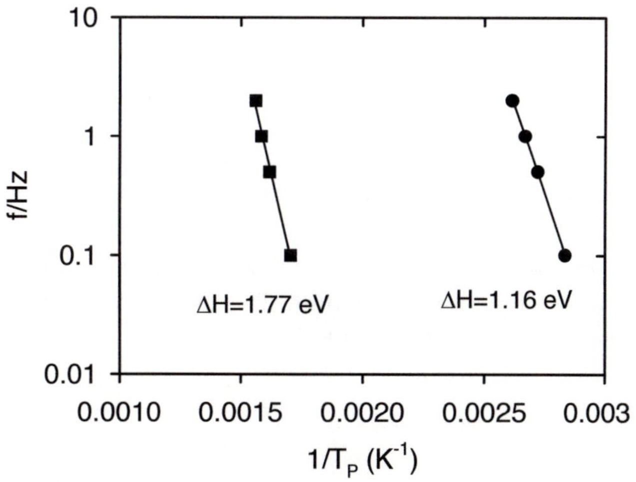

Figure 28. Arrhenius plots estimated for both peaks.

The internal friction in a material is a result of internal processes, which occurs during alternating stress cycles imposed on it and these processes originate from the interactions among structural components in the material. The presence of point, line, and planar defects within a stressed material often causes the internal friction to occur because of the atomic movement, rearrangement, or realignment of the defects under the application of a stress. These techniques can be used to characterize types of structure defects.

Figure 27 shows the temperature dependence of the internal friction IF=tanφ (Nowick & Berry. 1972) measured at 0.1 Hz during heating (heating rate 1 K/min.). Two peaks showing different intensities have been found: the weak low temperature peak (P1) and the more intense high temperature peak (P2). It can be seen that both peaks shift to a higher temperatures with increasing frequency, indicating that they may be relaxations peaks.

f (Hz)	P1 (°C)	P2 (°C)
0.1	79.6	313.2
0.5	94.7	344.5
1	103.3	357.9
2	109.5	367.2

Table 7. Temperatures of the relaxation peaks estimated for various frequencies.

The curves introduced in Fig. 27 were measured during one heating course in the multi-frequent mode. The internal friction peaks are assumed to be imposed by a background IF_b expressed by

$$IF_b = A + B \exp\left(-C_b / kT\right), \tag{14}$$

where k is the Boltzmann constant, T is the absolute temperature, A, B, and C_b are constants. After subtracting the background by using a fitting program PeakFit, the peak temperature T_P was estimated for the both relaxation peaks. The temperatures of both peaks estimated for various frequencies are given in Table 7. The peak widths are broader than that for a Debye peak, characterised by a single relaxation time. The internal friction peak appears at the condition $\omega\tau = 1$ (Nowick & Berry. 1972), with

$$\omega t = \omega t_0 \exp\left(DH / kT\right), \tag{15}$$

where ω is the angular frequency ($= 2\pi f$, f is the measuring frequency), τ is the mean relaxation time, τ_0 is the pre-exponential factor, and ΔH is the mean activation enthalpy.

http://dx.doi.org/10.5772/57454

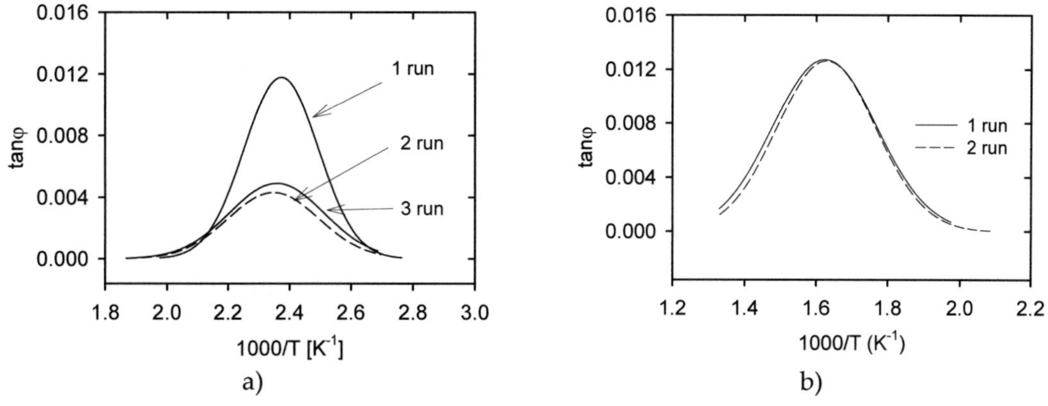

Figure 29. Background subtracted low (a) and high (b) temperature peaks after a prestraining of 2.55%.

4.2.1. Low temperature peak

Figure 28 shows the semilogarithmic plot of the frequency versus reciprocal value of peaks temperatures T_P (estimated for both peaks) so called Arrhenius plots. From the slope and intercept of the straight line, the mean activation enthalpy for the low temperature peak ΔH has been obtained and its value is 1.16±0.05 eV.

As far as the low temperature peak may have a dislocation nature, prestraining of the sample at low temperature should influence the height of the peak. Therefore, the samples were deformed by rolling. Deformation by rolling at room temperature ($\varepsilon \approx 2.55\%$) increases the strength of the low temperature peak, as it is obvious from Figure 29a (the peak was extracted from the internal friction spectrum subtracting the background damping). However, the peak is stronger only in the first run after prestraining. In the second and third run, the height of the peak is approximately the same as in the as-prepared state, and the height is lower than in the first run.

The strain amplitude dependent component of the logarithmic decrement (Figure 25a) indicates free dislocation loops in the material. IF peaks estimated in the second and third heating and cooling runs are practically the same. Beside this original dislocation population, the prestraining of the sample produced new dislocations which were practically completely recovered during the heating course of measurements. Owing to small grain size the grain boundaries were very probably the sources as well as sinks of the newly created dislocations. As mentioned above, the dominant deformation mode in magnesium is the basal slip; glide of <a> dislocations. The secondary conservative slip may be realised by motion of <a> dislocations on prismatic and pyramidal planes of the first kind. Couret & Caillard (1984a,b) studied by TEM prismatic glide in magnesium over a wide temperature range. They have reported that the screw dislocations with the Burgers vector of $1/3[11\bar{2}0]$ are able to glide on prismatic planes and their mobility is much lower than the mobility of edge dislocations. The deformation is controlled by the thermally activated glide of those screw dislocation segments. A single controlling mechanism has been identified as the Friedel-Escaig mechanism. This mechanism assumes a dissociated dislocation on a compact plane (0001) that joints together along a critical length L_r producing double kinks on non-compact plane. Dislocations in very small grains are pinned at the grain boundaries and this is the reason why this dislocation population is so stable. On the other hand, the newly created dislocations during cold rolling are only slightly pinned and they are recoverable during heating part of the measurement cycle.

4.2.2. High temperature peak

The high temperature peak was observed at a temperature of ≈ 345 °C (0.5Hz). The cold prestraining of the sample did not affect the strength of the high temperature peak, as it can be seen from Fig. 29b. Position and height of the peak are very stable during heating as well as cooling. The activation energy was obtained from the frequency dependence of the peak temperature (Arrhenius plot) to be 1.77±0.05 eV (see Figure 28).

The presence of the high temperature peak in magnesium has been reported by Kê (Kê, 1947). He described this peak as being related to the grain boundary relaxation.

Grain boundaries in materials with the small grain size contain a dense network of overlapping grain boundary dislocations. Grain boundary sliding is realized by the slip and climb (providing a maintenance of vacancy sources and sinks) of the grain boundary dislocations. Since both modes of the grain boundary dislocation motion must occur simultaneously, the slower one will control the grain boundary sliding. The climb mode involves jog formation and grain boundary diffusion and both modes of the grain boundary dislocation movement may be affected by the interaction of the grain boundary dislocations with impurities segregated in grain boundaries. According to Lakki et al. (1998) the activation energy for such process can be expressed as $\Delta H = \Delta H_j + \Delta H_{GB}$, where ΔH_j and ΔH_{GB} are the activation enthalpies for the jog formation and the grain boundary diffusion, respectively. The activation enthalpy for grain boundary diffusion in the coarse grained magnesium is 0.95 eV (Frost & Ashby, 1982). Comparing with the estimated activation enthalpy value of 1.77 eV, the possible jog formation energy should be approximately 0.8 eV, which seems to be a reasonable value. Taking into consideration that the surface of grains was passivated during the milling by 1% of oxygen in a protective argon atmosphere; therefore, MgO particles in the grain boundaries may influence the grain boundary diffusion characteristics.

Acknowledgements

Z.T. and P.L. are grateful for the financial support to the Czech Grant Agency under the contract P108/12/J018.

Author details

Zuzanka Trojanová[1], Pavel Lukáč[1], Zoltán Száraz[2] and Zdeněk Drozd[1]

1 Charles University in Prague, Czech Republic

2 The University of Manchester, UK

References

[1] Andersson, P., Cáceres, C.H. & Koike, J. (2003) Hall-Petch parameters for tension and compression in cast Mg. *Mater. Sci. Forum* 419–42, 123–128. ISSN: 1662-9752

[2] Chawla, K.K. (1991) Metal matrix composites. *VCH Weinheim,* ISBN: 03872-85679

[3] Arsenault, R.J. & Shi, N. (1986). Dislocation generation due to differences between the coefficients of thermal expansion. *Mater. Sci. Eng.*, 81, 175-187. ISSN: 0254-0584.

[4] Choi, H.J., Kim, Y., Shin, J.H. & Bae, D.H. (2010) Deformation behavior of magnesium in the grain size spectrum from nano- to micrometer. *Mater. Sci. Eng.* A 527, 1565–1570. ISSN: 0254-0584.

[5] Couret, A. & Caillard, D. (1985a) An in situ study of prismatic glide in magnesium—I. The rate controlling mechanism. *Acta Metall.* 33 1447-1454. ISSN: 0921-5093

[6] Courret, A. & Caillard, D. (1985b) An *in situ* study of prismatic glide in magnesium—II. Microscopic activation parameters, *Acta Metall.* 33, 1455-1462. ISSN: 0921-5093

[7] Dieringa, H. (2011) Properties of magnesium alloys reinforced with nanoparticles and carbon nanotubes: a review. *J. Mater. Sci.* 46, 289–306. ISSN: 0022-2461

[8] Dragomir, I.C. & Ungár, T. (2002) The contrast factors of dislocations in the hexagonal crystal system. *J. Appl. Cryst.* 35, 556-564.

[9] Dunand, D.C. & Mortensen, A. (1991) On plastic relaxation of thermal stresses in reinforced metals. *Acta Metall. Mater.* 39, 127-139. ISSN: 0921-5093

[10] Estrin, Y. & Vinogradov A. (2013) *Acta Mater.* 61, 782-817. ISSN: 0921-5093

[11] Ferkel, H. (2003) Properties of graphite-strengthened magnesium. *Adv. Eng. Mater.* 5, 886-889. ISSN: 1527-2648

[12] Ferkel, H. & Mordike, B.L. (2001) Magnesium strengthened by SiC nanoparticles. *Mater. Sci. Eng. A* 298, 193–199. ISSN: 0254-0584

[13] Frost, H.J. & Ashby, M. F., Deformation-mechanism maps. Pergamon Press, Oxford, UK, 1982, 44. ISBN 0-08-029337-9

[14] Granato, A.V. & Lücke, K. (1981) Temperature dependence of amplitude-dependent dislocation damping` *J. Appl. Phys.*, 52, 7136-7142. ISSN 0021-8979

[15] Ion, S.E., Humphrey, F.J. & White, S.H. (1982) Dynamic recrystallization and the development of microstructure during hot deformation of magnesium. *Acta Met.* 30, 1909-1919. ISSN: 0921-5093

[16] Kê, T.S. (1947) Experimental evidence of the viscous behaviour of grain boundaries in metals *Phys. Rev.* 71, 533-546. 1050-2947

[17] Koike, T., Kobayashi, T., Mukai, T., Watanabe, T.H., Suzuki, M., Maruyama, K. & Higashi, K. (2003) The activity of non-basal slip systems and dynamic recovery at room

temperature in fine-grained AZ31B magnesium alloys. *Acta Mater.* 51, 2055–2065. ISSN: 0921-5093

[18] Kužel, R. & Klimanek, P. (1988) X ray diffraction line broadening due to dislocations in non-cubic materials. 2. The case of line anisotropy applied to hexagonal crystals. *J. Appl.Cryst.* 21, 363-368. ISSN: 0021-8898

[19] Kužel, R. (1998) DIFPATAN—Program for Powder Pattern Analysis. http://www.xray.cz/ecm-cd/soft/xray/indexdifp.html

[20] Lakki, A., Schaller, R., Carry, C. & Benoit, W. (1998) High temperature anelastic and viscoplastic deformation of fine-grained MgO-doped Al_2O_3`, *Acta Mater.*, 46, 689-700. ISSN: 0921-5093

[21] Langdon, T.G. (1994) A unified approach to grain boundary sliding and superplasticity. *Acta Metall. Mater.* 42, 2437-2443. ISSN: 0921-5093

[22] Lukáč, P., Trojanová, Z., Száraz, Z., Svoboda, M. & Ferbl, H. (2006) Deformation behaviour of ultrafine grained magnesium with 3 vol. % graphite. *Z. Metallkd.* 97, 344-349. ISSS: 1862-5282

[23] Mabuchi, M. & Higashi, K. (1999) On accommodation helper mechanisms for superplasticity in metal matrix composites. *Acta Mater.* 47, 1915-1922. ISSN: 0921-5093

[24] Mabuchi, M., Chino, Y., Iwasaki, H., Aizawa, T. & Higashi, K. (2001) The grain size and texture dependence of tensile properties in extruded Mg–9Al–1Zn. *Mater. Trans. 42,* 1182–1189. ISSN: 1345-9678

[25] Máthis, K., Nyilas, K., Axt, A., Dragomir-Cernatescu, I., Ungár, T. & Lukáč, P. (2004) The evolution of non-basal dislocations as a function of deformation temperature in pure magnesium determined by X-ray diffraction, *Acta Mater.*, 52, 2889–2894. ISSN: 0921-5093

[26] Máthis, K., Čapek, J., Zdražilová, Z. & Trojanová, Z. (2011) Investigation of tension–compression asymmetry of magnesium by use of the acoustic emission technique. *Mater. Sci. Eng. A* 528, 5904–5907. ISSN: 0254-0584

[27] Mohamed, F.A. (2003) A dislocation model for the minimum grain size obtainable by milling. *Acta Mater.* 51, 4107–4119. ISSN: 0921-5093

[28] Moll, F., Chmelík, F., Lukáč, P., Mordike, B.L. & Kainer, K.U. (2000) Creep behaviour of a QE22-SiC particle reinforced composite investigated by acoustic emission and scanning electron microscopy. *Mater. Sci. Eng. A* 291, 246-249. ISSN: 0254-0584

[29] Naser, J., Riehemann, W. & Ferkel, H. (1997) Dispersion hardening of metals by nanoscaled ceramic powders. *Mater. Sci. Eng.* A 234–236, 467–469. ISSN: 0254-0584

[30] Nayeb-Hashemi, A.A., Clark, J.B., & Pelton A.D. The Li-Mg (Lithium-Magnesium) system, *Bulletin of Alloy Phase Diagrams,* 5, 365-374, 1984. ISSN: 1054-9714

[31] Nes, E., Holmedal, B., Evangelista, E. & Marthinsen, K. (2005) Modelling grain boundary strengthening in ultra-fine aluminium alloys`, *Mater. Sci. Eng. A*, 410–411, 178–182. ISSN: 0254-0584

[32] Nowick, A.S. & Berry, B.S. (1972) *Anelastic Relaxations in Crystalline Solids* Academic Press, New York/London. ISBN: 70-154378

[33] Száraz, Z.. Trojanová, Z., Cabbibo, M., & Evangelista, E. (2007) Strengthening in a WE54 magnesium alloy containing SiC particles. *Mater. Sci. Eng. A*, 462, 225-229. ISSN: 0254-0584

[34] Thostenson, E.T., Li, C. & Chou, T.-W. (2005) Nanocomposites in context. *Compos. Sci. Technol.* 65, 491–516. ISSN: 0266-3538

[35] Trojanová, Z., Lukáč, P., Ferkel, H., Mordike, B.L. & Riehemann, W. (1997) Stability of microstructure in magnesium reinforced by nanoscaled alumina particles. *Mater. Sci. Eng. A* 234-236, 798-801. ISSN: 0254-0584

[36] Trojanová, Z, Száraz, Z., Mielczarek, A, Lukáč, P. & Riehemann, W. (2008) Plastic and fatigue behaviour of ultrafine-grained magnesium. *Mater. Sci. Eng. A* 483–484, 477-480. ISSN: 0254-0584

[37] Trojanová, Z., Máthis, K., Lukáč, P., Németh, G. & Chmelík, F. (2011) Internal stress and thermally activated dislocation motion in an AZ63 magnesium alloy. *Mater. Chem. Phys.* 130, 1146-1150. ISSN: 0254-0584.

[38] Ungár, T. & Borbely, A. (1996) The effect of dislocation contrast on X-ray line broadening: a new approach to line profile analysis. *Appl. Phys. Let.*; 69, 3173-3175. ISSN: 0003-6951

[39] Wilkens, M. (1970) `The determination of density and distribution of dislocations in-deformed single crystals from broadened X-ray diffraction profiles, *phys. stat. sol. (a)*, 2, 359-370. ISSN: 1862-6319

[40] Yoo, M.H., Agnew, S.R., Morris, J.R. & Ho, K.M. (2001) Non-basal slip systems in HCP metals and alloys: source mechanisms. *Mater. Sci. Eng. A*, 319-321, 87-92 ISSN: 0254-0584

[41] Zhang, J., Perez, R.J., Wong, C.R. & Lavernia, E.J. (1994a) Effects of secondary phases on the damping behaviour of metals, alloys and metal matrix composites. *Mater. Sci. Eng.* R13, 325-390. ISSN: 0254-0584

[42] Zhang, J., Perez, R.J. & Lavernia, E.J. (1994b) Effect of SiC and graphite particulates on the damping behavior of metal matrix composites. *Acta Metal. Mater.*, 42, 395-409. ISSN: 0921-5093

Chapter 9

Heuristic Analysis Application for Magnesium Alloys Properties Improvement

Tomasz Tański, Krzysztof Labisz and
Anna Dobrzańska-Danikiewicz

Additional information is available at the end of the chapter

http://dx.doi.org/10.5772/57587

1. Introduction

The modern development of technology makes it necessary to look for new design solutions to improve efficiency and product quality, to minimize size and to increase the reliability and dimensional stability under service conditions. The choice of material is always preceded by an analysis of a number of factors including the requirements like: mechanical, physical and chemical, design, environmental, cost related, availability and weight. The problem of a too high mass verifies strongly the applicability of particular groups of materials. Therefore, over the past few decades, occurs a significant increase in demand for materials with low density and relatively high strength alloys such as titanium, aluminium and magnesium. In this group of materials special attention is devoted to magnesium alloys, which in addition to low densities have also other advantages, such as good damping capacity, the best among all currently known structural materials, high dimensional stability, good fluidity, low shrinkage of casts, a desirable combination of low density and high strength, and it can be used for machine parts operating at temperatures up to 300°C as well as relatively ease recyclability. The high damping capacity and relatively low weight allows to use magnesium alloys for production of fast moving parts where sudden changes in speed occurs. However, magnesium is an undisputed advantage especially its low density (1.7 g/cm^3), which is about 35% less than the density of aluminium (2.7 g/cm^3) and more than four times lower than the density of steel (7.86 g/cm3).

2. Investigation method

Investigations described in this chapter concerns first of all investigation of structure and properties of the described engineering materials, in the state after heat and surface treatment, including the selected group of magnesium alloys, aluminium alloys and steel, as well as heuristic analysis using the methodology of integrated computer prediction, enabling the demarcation of predicted development trends of the analysed groups of materials and technologies of surface treatment and lead to the determination of their strategic development in comparison with other groups of materials and surface engineering technologies. In the initial planning stage of research scope in this investigations, in the context of choosing specific research topics with a comprehensive range of academic disciplines or specialties an innovative analysis was used to select the technology that uses the complex dendro-chronological evaluation matrix. Furthermore, in order to confirm the appropriateness of the proper selection of the investigated substrate material like the described in this paper Mg-Al-Zn cast magnesium alloys, there was performed an analysis using the dendrological selection matrix, considering only three groups of most commonly used construction materials, namely steel, aluminium and magnesium alloys.

In addition, the presented "hard" investigation results concerns the characteristics of synergistic effect of the surface heat treatment on the structure and properties of the cast of magnesium alloys, aluminium alloys and steel. Surface treatment of the analysed materials was made using laser surface treatment methods, including, in particular, laser alloying and feeding of hard ceramic particles in the surface of the material to allow formation of quasi-MMCs composite structure (Metal Matrix Composites) on the surface. Laser surface feeding and alloying was conducted by remelting of magnesium, aluminium alloys and steel surface with hard carbide and oxide particles. The feeding and alloying materials were BN, Si_3N_4, ZrO_2, Al_2O_3, NbC, TaC, VC, WC, TiC, SiC powders. The powders was supplied by side injection rate of 6±1 g/min. The laser feeding was performed by high power laser diode HPDL Rofin DL 020 using the argon shielding gas. Argon was used during laser remelting to prevent oxidation of the coating and the substrate. The process parameters during the present investigation were: laser power-1.2, 1.6, 2.0 kW, scan rate – 0.5-1.0 m/min.

The observations of the investigated materials have been made on the light microscope LEICA MEF4A as well as on the scanning electron microscope (SEM) ZEISS SUPRA 35 and on the JEOL JEM 3010 transmission electron microscope (TEM), at the accelerating voltage of 300 kV.

Hardness tests were made using Zwick ZHR 4150 TK hardness tester in the HRF scale. Microhardness of the cross section of the laser surface remelted layer was measured on Fully-Automatic Microhardness Testing System with a loading time of 15 s and the testing load of 100 g.

http://dx.doi.org/10.5772/57587

3. Investigation results

3.1. First part – Heuristic analysis

In the first step of the heuristic material investigation there should be distinguished among all the available constructional materials and technologies several homogeneous groups of materials for the reason to undertake the planned experimental and comparative analysis. The impact of external positive factors (opportunities) and negative factors (difficulties) on the specific technologies analysed was evaluated using the metrological matrix of environment influence. In consistency with the concept proposed, each of the technologies assessed by the experts for their position against the macro-and microenvironment is placed in one of the quarters of this matrix. A universal scale of relative states, being a single-pole positive scale without zero, where 1 is a minimum value and 10 an extraordinarily high value, was used in the research undertaken. The dendrological matrix of values allows in a relatively easy way to visualize the results of the assessment of individual groups of materials and in a next steps of processing technologies in terms of potential, which is the real objective value of the studied area, the so-called hard values, tangible qualities as well as attractiveness, which reflects the subjective perception of its potential users,the so-called soft properties. The potential of a given group presented on the horizontal axis is the result of multi-criteria analysis carried out on the basis of expertise and extensive knowledge of literature data, taking into account the appropriate proportions of the potential: creativeness, application, quality, and technical development. The vertical axis shows the importance of the attractiveness of the group, which is average mean value of expert and literature data evaluation in the field of research carried out on the basis of specific criteria corresponding to the economic attractiveness, cultural, economic, scientific and systemic. The technologies analysed were first evaluated by key experts for their attractiveness and potential and the result obtained was entered into one of the quarters of dendrological matrix of technology value. The wide-stretching oak is the most promising quarter guaranteeing the future success encompassing the technologies with a high potential and attractiveness. The soaring cypress includes very attractive technologies with a limited potential and rooted dwarf mountain pine symbolises technologies with a high potential and limited attractiveness likely to ensure a strong position for the technology if an adequate strategy is employed. The least promising technologies are entered in the quarter called the quaking aspen with their success being either unlikely or impossible. The evaluation of the technology value made has revealed that all the groups of technologies are characterised by a high potential and attractiveness, thus classified to the most promising quarter called the wide-stretching oak. (Fig. 1).

Investigations of processing technology of the used alloys were tightened to the area of surface engineering only, because there exist a wide variety of available types of coatings and methods for forming structure and properties of the surface layer of engineering materials, including magnesium alloys. The investigations allow to determine precise and complete the way to design preferably summarized core material properties as well as the surface layer of the produced element.

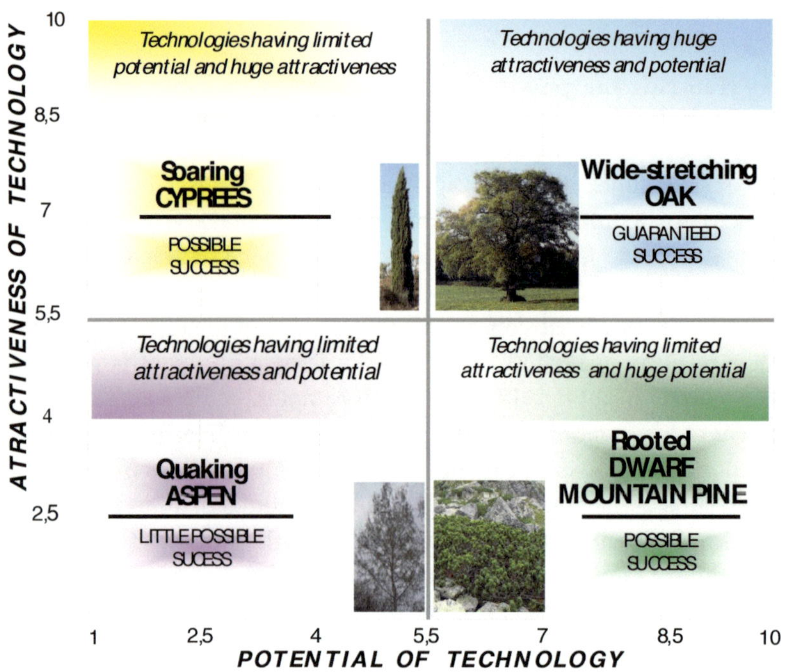

Figure 1. Dendrological matrix of technology value; approach presentation [by A.D. Dobrzańska-Danikiewicz]

For determination of the value of each group of materials, and in the second step of their surface treatment technology, there was applied the weighted scoring method for a comparative assessment used for appropriate qualifications for the specific groups of materials and technologies in the context of their mutual relationship. It should be noted that this technique involves the application of the relativistic principle of evaluation criteria, namely the recognition of differences and existing rules between the different research areas, assumed a specific set of criteria, acting as a selective filter evaluating positively or negatively the investigated object. The weighted scoring method enables the use of multi-criteria evaluation of areas which acts as the perfect model of synthetic evaluation of specific areas, using an interval scale.

Detailed criteria for evaluation of the attractiveness and potential of investigated areas-concerning materials and technologies-investigated in this work are presented in Tables 1 and 3. The listed criteria assigned specific importance, used for following calculation of the weighted values and their sum, which is the basis for comparative analysis, as shown in Tables 2 and 4. Then the multi-criteria results are visualized using a matrix of dendrological values for the investigated area (Figs. 2 and 3). The analysis of preferences clearly indicates, that the highest weighted evaluation, taking into account the criteria in each of the group of materials, are determined for magnesium alloys, which meet a wide range of demands and expectations of modern engineering materials imposed by the ever-changing environment, marked in Fig. 2 as M1 and coloured green. The investigations confirms also, that the aluminium alloys are characterized by a high level of attractiveness.

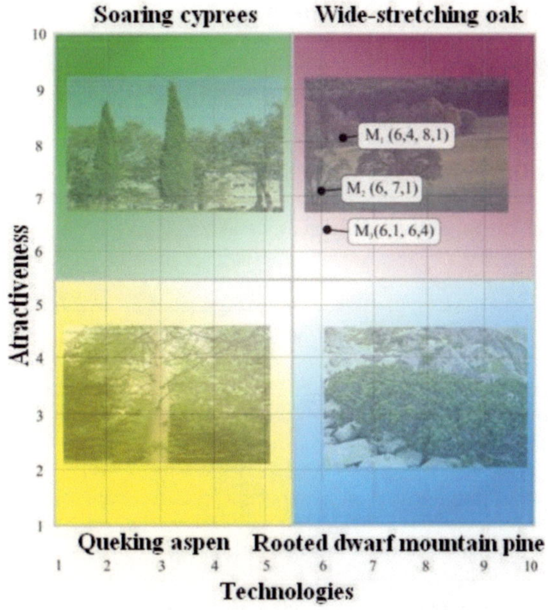

Figure 2. Dendrochronology matrix of values presenting the placement of evaluated groups of materials investigated using computer integrated forecasting of the development

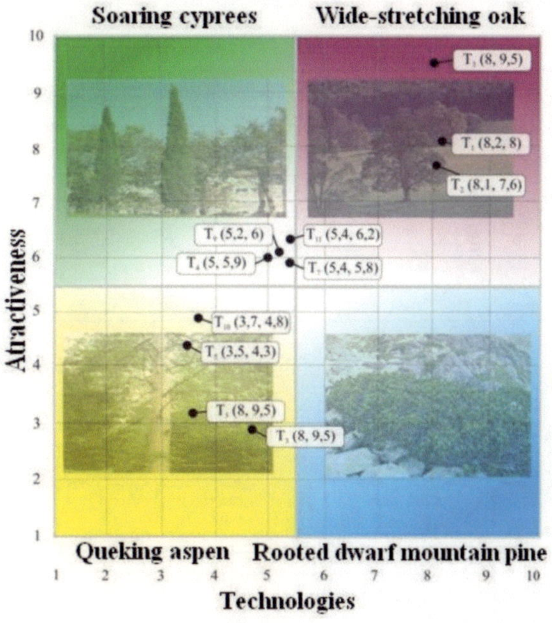

Figure 3. Dendrochronology matrix of values presenting the placement of evaluated groups of technologies investigated using computer integrated forecasting of the development

No.	Potential	Importance
Criterion 1	Low specific gravity of the material	0,4
Criterion 2	Ability for vibration damping	0,1
Criterion 3	Good castability	0,1
Criterion 4	High mechanical properties and good weldability	0,2
Criterion 5	High corrosion resistance	0,2
No.	**Attractiveness**	**Importance**
Criterion 1	Availability of the half-finished product (raw material, chemical element)	0,4
Criterion 2	Recycling possibility	0,1
Criterion 3	Low production costs and the final price of the produced material	0,2
Criterion 4	Low level of complexity of the production technology	0,1
Criterion 5	The variety of technologies for semi-fabrication of the material and the wide choice of processing technologies possible to implement (heat treatment, plastic deformation, surface treatment)	0,2

Table 1. Particular criteria for evaluation of the potential and attractiveness of material groups using for material engineering-heuristic investigations

Designation	Group of material	Potential						Attractiveness					
		Criterion 1	Criterion 2	Criterion 3	Criterion 4	Criterion 5	weighted average	Criterion 1	Criterion 2	Criterion 3	Criterion 4	Criterion 5	weighted average
M_1	Magnesium alloys	4	0,8	0,8	0,6	0,2	**6,4**	4	0,9	1,2	0,4	1,6	**8,1**
M_2	Aluminium alloys	2,8	0,5	0,7	0,8	1,2	6	2,8	0,7	1,2	0,8	1,6	7,1
M_3	Steels	1,6	0,4	0,5	1,8	1,8	6,1	2	0,4	1,6	0,8	1,6	6,4

Table 2. Multi-criterion analysis of the investigated materials groups using engineering-heuristic methods

The highest evaluation are replaced in the group of technologies (the quadrant of a large oak) by laser technologies and PVD/CVD methods, indicated in Fig. 3 as green coloured T1, T2 and

T3 points-selected as the most suitable surface treatment technologies for magnesium alloys, far exceeding the other analysed coated methods. Other techniques were evaluated in the area of the quadrant of trembling aspen or soaring cypress which may indicate, in the case of technology in the area of a trembling aspen a slow displacement of these methods by newer more environmentally friendly and versatile technology.

However, in case of technologies, which are placed in the area of soaring cypress there it is a need for further research and investments enabling them to develop. The only exception to the assumptions may be layering paint technologies that typically are located in the square of rooted dwarf mountain pine, in relation to painting. This technology is unattractive, but widely used due to its simple mechanism for layering and low application cost. However, according to the established criteria concerning the potential and attractiveness for future application specific to the substrate materials (magnesium alloy); the painting techniques are qualified for the quarter trembling aspen. The selected technologies-T1, T2 and T3 were characterised in detail in the following description.

No.	Potential	Importance
Criterion 1	The possibility to obtain complex coatings (complex, multi-layered, multi-phase, gradient, composite, metastable, nanocrystalline)	0,3
Criterion 2	Large choice of coating material; wide range of deposited coatings properties	0,2
Criterion 3	The possibility to obtain hard surface layers with special protective properties (corrosion resistant, wear resistant)	0,2
Criterion 4	Layers with good adhesion to the substrate	0,2
Criterion 5	The possibility to obtain in a single process step graded layers with defined chemical composition and structure	0,1
No.	**Attractiveness**	**Importance**
Criterion 1	Environmental friendly deposition process (no harmful products of chemical reactions and the need for their recycling)	0,2
Criterion 2	Possibility to produce coatings with properties impossible to obtain with other methods	0,3
Criterion 3	Wide range of further development of the technology	0,2
Criterion 4	Possibility of full automation of the manufacturing process of coating	0,1
Criterion 5	Necessity for high precision of the deposition processes	0,2

Table 3. Detailed criteria for the assessment of the potential and attractiveness of the group of investigated material science and heuristic technologies

Designation	Technology group	Potential						Attractiveness					
		Criterion 1	Criterion 2	Criterion 3	Criterion 4	Criterion 5	weighted average	Criterion 1	Criterion 2	Criterion 3	Criterion 4	Criterion 5	weighted average
T_1	PVD techniques, including also the Cathodic Arc Deposition	2,7	1,6	1,8	1,2	0,9	**8,2**	1,8	2,7	1,6	0,9	1	**8**
T_2	CVD techniques, including the PACVD	2,7	1,6	1,8	1,2	0,8	**8,1**	1,4	2,7	1,6	0,9	1	**7,6**
T_3	Laser feeding/ alloying/remelting	2,7	1	1,8	2	0,5	**8**	1,8	2,7	2	1	2	**9,5**
T_4	Thermal spraying	1,5	1	1,4	0,6	0,5	5	1,2	1,8	0,6	0,9	1,4	5,9
T_5	Anodisation	0,6	0,2	1,2	1,2	0,3	3,5	1,2	1,5	0,2	0,4	1	4,3
T_6	Galvanic technologies	0,9	1	1,2	1,4	0,2	4,7	0,4	0,9	0,2	0,9	0,4	2,8
T_7	Laser ablation-PLD	1,5	1	1	1,4	0,5	5,4	1,6	1,2	1,2	0,6	1,2	5,8
T_8	Painting coating	0,3	0,8	1	1,4	0,1	3,6	0,6	0,6	0,2	0,9	0,8	3,1
T_9	Ion implantation	1,5	1	1	1,2	0,5	5,2	1,4	1,8	1,2	0,6	1	6
T_{10}	Arc nitriding	0,6	0,4	1,2	1,2	0,3	3,7	0,8	1,8	0,6	0,8	0,8	4,8
T_{11}	Hybrid technologies (multiplex)	1,5	1	1,2	1,2	0,5	5,4	1	2,7	1,2	0,5	0,8	6,2

Table 4. The results of multi-criteria analysis of the investigated groups of technologies

3.2. Second part – Material engineering investigations

The developed technologies focusing on hybrid surface layers, so-called quasi-composite MMCs structures (characterized by phase composition gradient, chemical composition and functional gradient) in the process of laser alloying and/or feeding with ceramic particles into the surface of the treated materials provide a complete and comprehensive solution for modelling of engineering materials. Currently, the concept of functional lightweight materials is a priority and the most investigated worldwide field of material science and engineering concerning the production and processing of new developed engineering materials. Previous studies about the effects of laser beam effect on various materials, including magnesium alloys as well as tool steels, based on the authors own long-term investigation, summarizing the experience of laser surface treatment reveal that, there are chemical composition and structure changes which are different from those occurring during conventional heat treatment. This causes, that the laser treated elements shows relatively high hardness (Fig. 4) and thermal

http://dx.doi.org/10.5772/57587

fatigue (Fig. 5), especially for VC-alloyed steel surface (Figs. 6, 7 and 10, 11) compared to the traditionally treated materials.

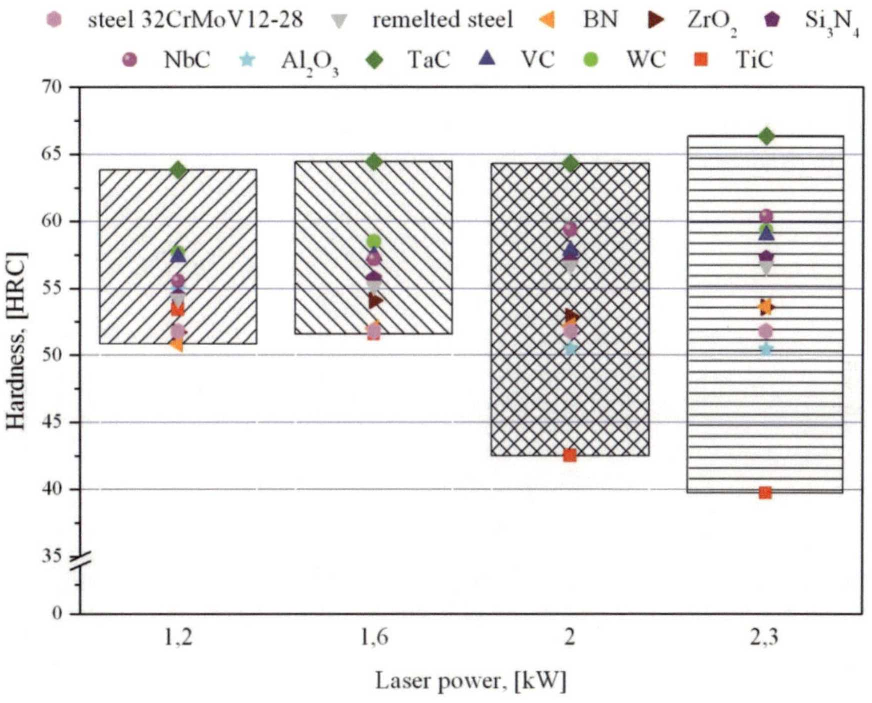

Figure 4. Hardness measurement results of the 32CrMoV12-28 steel alloyed with ceramic powders

Similar relationship can be found in case of Si_3N_4 powder alloying, with lower hardness and surface roughness compared to the VC alloying variant (Figs. 8, 9 and 12, 13).

The performed investigations of the alloyed hot work tool steel 32CrMoV12-28 show a clear effect of the applied ceramic powder used for alloying (Figs. 8, 9). It can be also clearly recognised the influence of the used laser power in the range of 1.2; 1.6; 2.0 and 2.3 kW on the shape and thickness as well as the particle distribution of the remelted material. It can be seen that with the increasing laser power the distribution of the remelted metal in the steel substrate increases. Microstructure presented on Figs. 8, 9 shows a dendritic structure in the remelted area. There are also VC particles present distributed in the matrix confirmed by the transmission electron microscope using the selected area diffraction pattern. There is also a clear relationship between the employed laser power and the dendrite size, namely with increasing laser power the dendrites are larger, this relationship is valid for the Si_3N_4 powder alloying.

In general the hot work tool steel has a ferritic structure with homogeny distributed carbides in the metal matrix in the annealed state. In areas, which are between the solid and molten state dendritic structure with large dendrites can be found in case of Si_3N_4 powder (Fig. 14) or a fine grained structure in case of VC alloying (Fig. 15). The EDS point wise chemical composition analysis confirms the presence of carbide ceramic particles in the matrix in form of big conglomerates. The required hardenability for this tool steel was achieving after a suitable tempering time, which assures melting of the alloying carbides in the austenite. The structural investigations carried out using the high power diode laser allows to compare the surface layer as well as the shape and depth of the remelting area. It was noticed that the depth of remelting area grows together with the increasing laser power (Fig. 16).

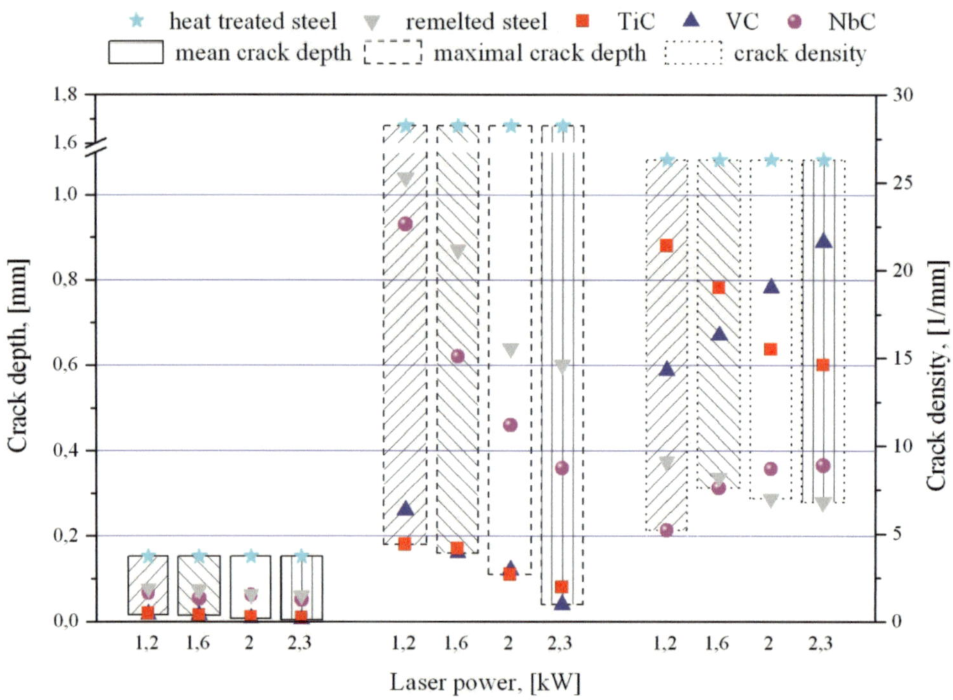

Figure 5. Influence of laser power and the type of alloyed surface layer of the 32CrMoV12-28 steel on mean crack depth, maximal crack depth and crack density occurred during the thermal fatigue test

It was found, that in case of VC powder the difference of the remelted area thickness is several times higher in case of 2.3 kW power compared to 1.2 kW laser power. Also for the same laser power (2.3 kW) the surface layer thickness increases from 1.9 mm for WC powder to 2.2 mm for TiC powder and to 2.3 mm in case of the VC powder (Fig. 16).

http://dx.doi.org/10.5772/57587

Figure 6. 32CrMoV12-28 steel alloyed with VC powder, laser power 1.2 kW

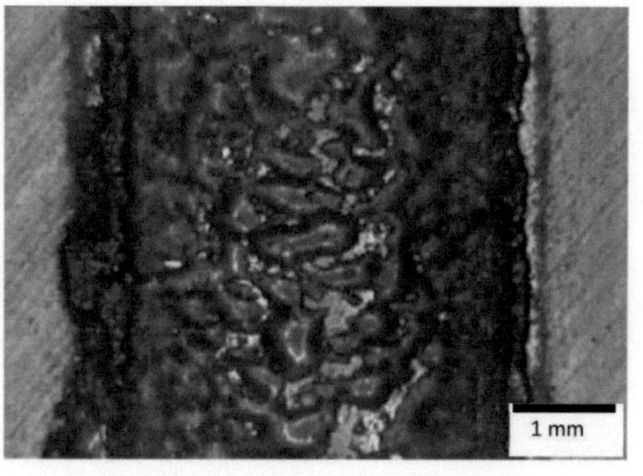

Figure 7. 32CrMoV12-28 steel alloyed with VC powder, laser power 2.3 kW

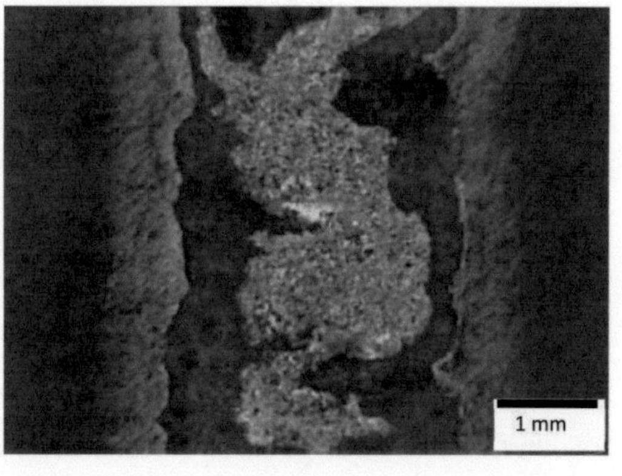

Figure 8. .32CrMoV12-28 steel alloyed with Si_3N_4 powder, laser power 1.2 kW

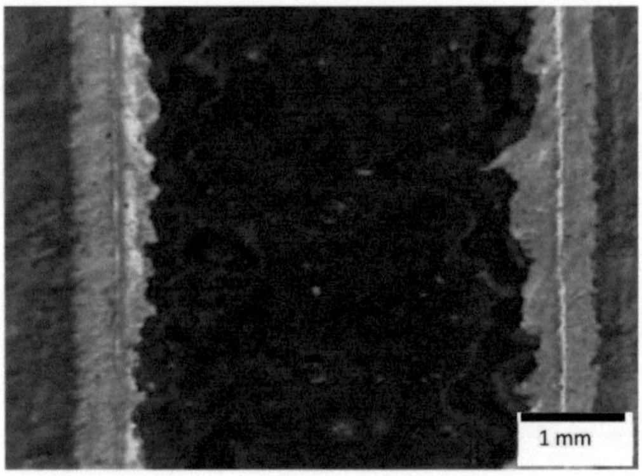

Figure 9. 32CrMoV12-28 steel alloyed with Si_3N_4 powder, laser power 2.3 kW

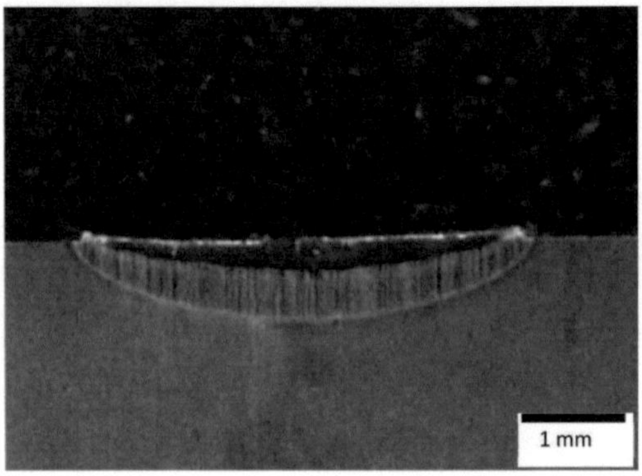

Figure 10. 32CrMoV12-28 steel alloyed with VC powder, laser power 1.2 kW

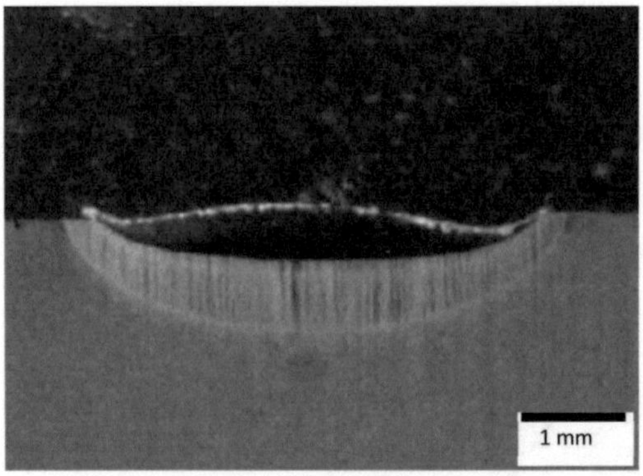

Figure 11. 32CrMoV12-28 steel alloyed with VC powder, laser power 1.2 kW

http://dx.doi.org/10.5772/57587

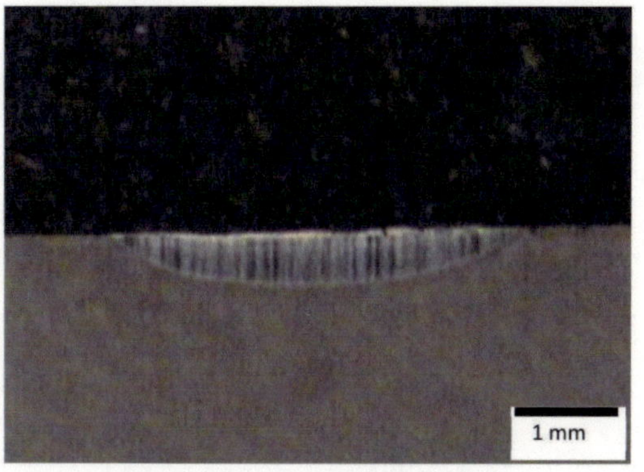

Figure 12. 32CrMoV12-28 steel alloyed with Si_3N_4 powder, laser power 1.2 kW

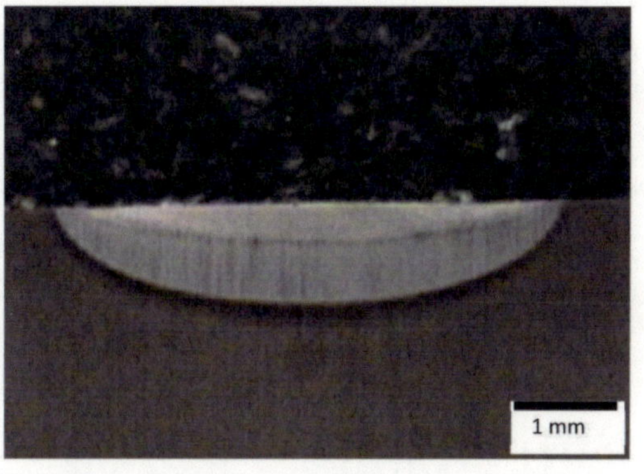

Figure 13. 32CrMoV12-28 steel alloyed with Si_3N_4 powder, laser power 2.3 kW

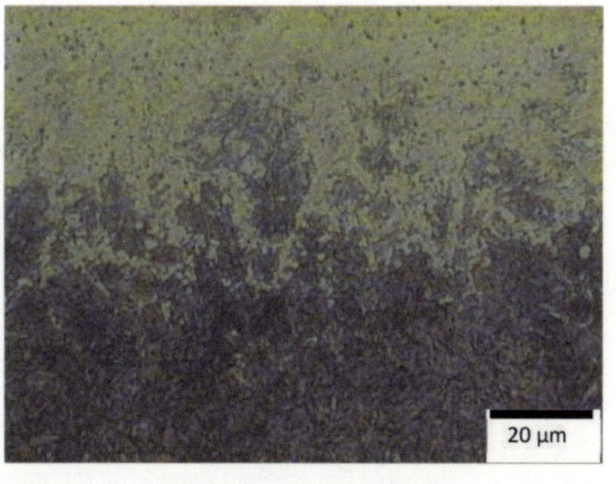

Figure 14. 32CrMoV12-28 steel alloyed with Si_3N_4 powder, laser power 1.2 kW

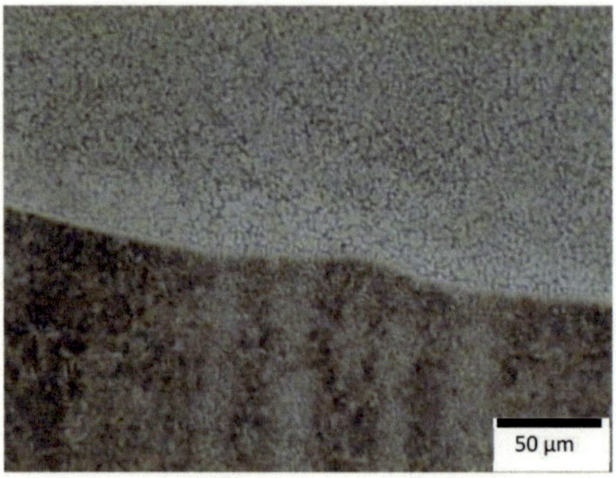

Figure 15. 32CrMoV12-28 steel alloyed with VC powder, 1.2 kW

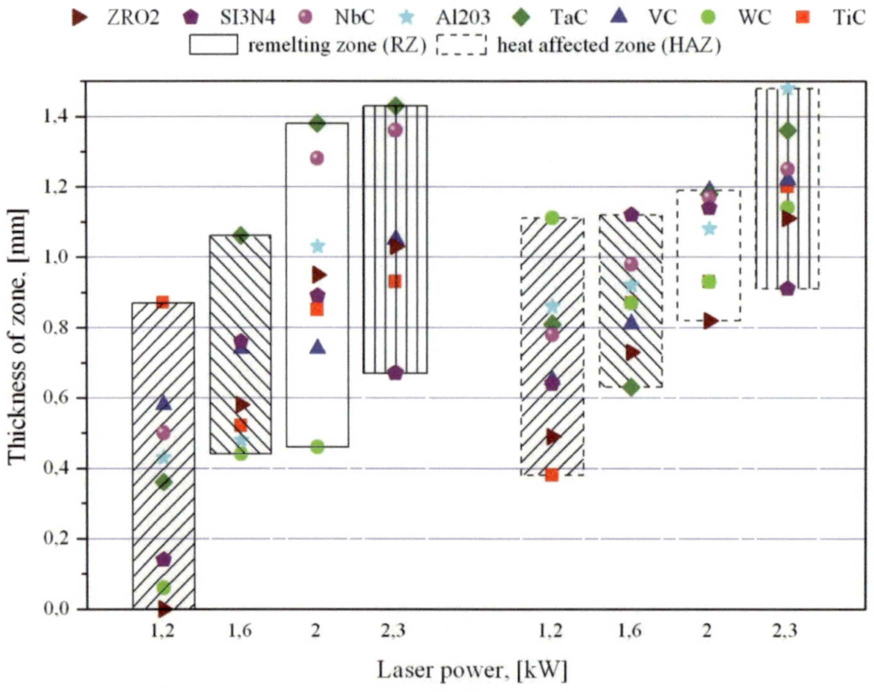

Figure 16. Influence of laser power on thickness of the remelted zone RZ, heat affected zone HAZ and the surface layer 32CrMoV12-28 steel after laser alloyed.

http://dx.doi.org/10.5772/57587

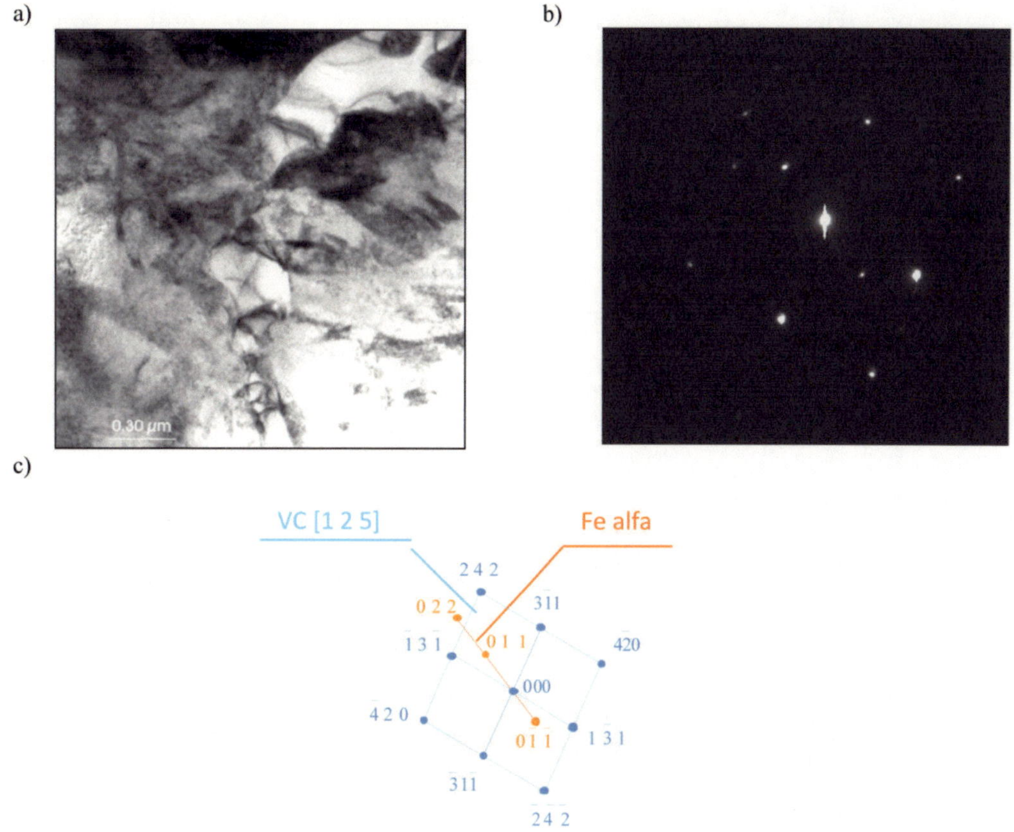

Figure 17. Thin foil structure of the 32CrMoV12-28 steel after VC alloying, laser power– 2,0 kW, a) bright field, b) diffraction pattern of the VC particle revealed in Fig. 17a, c) solution of the diffraction pattern from Fig. 17c

Similar relationships were found in case of magnesium alloys, where the used WC, VC, SiC, Al_2O_3 and TiC powder particles are present in the laser treated magnesium surface. Effect of laser feeding conditions, namely: laser power, feeding speed, type of the used ceramic powder and the applied substrate on hardness and hardness increase of the surface layer of the cast magnesium samples were investigated using the Rockwell hardness method. The measured hardness of the surface was obtained in the range from 32.4 to 105.1 HRF (Fig. 18). As a result of the performed investigations it was found, that the highest hardness increase was observed in case of MCMgAl6Znl and MCMgAl3Znl magnesium cast alloys – as materials with low concentration (<6%) of aluminium, which is mainly responsible for precipitation strengthening of the studied, laser treated alloys, fed with ceramic particles.

Also in this case the surface layer is relatively rough, compared to the non-treated material (Fig. 19, 20). After laser feeding, there was revealed-based on the performed metallographic investigations which were carried out on light microscope-the presence of several zones in the remelted surface layer of the cast magnesium alloys, with the thickness and the powder particles occurrence depending on the laser processing parameters and the used powder and substrate type (Figs. 21, 22). Starting from the top zone of the surface layer there occurs one zone rich in non-dissolved particles located on the surface of magnesium alloys, the second zone is the remelting area zone (RZ), with the thickness and shape directly depending on the applied laser power. Finally there occurs also the heat affected zone (HAZ) (Figs. 23, 24).

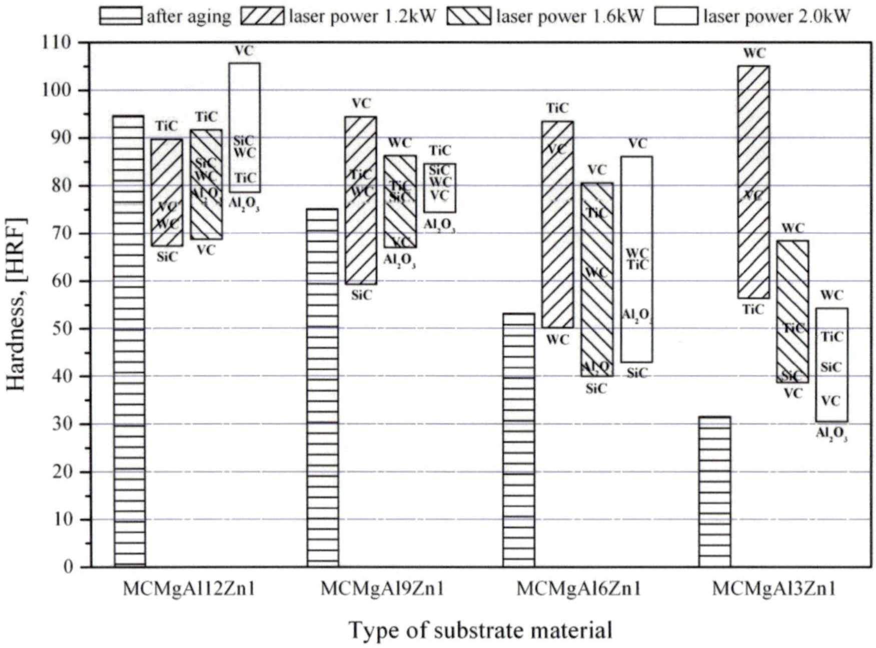

Figure 18. Hardness measurements results of cast Mg-Al-Zn magnesium alloys samples, after aging and laser feeding

Figure 19. Surface layer of the MCMgAl9Zn1 alloy after WC powder feeding, laser power 2.0 kW, laser scan rate 0.5 m/min

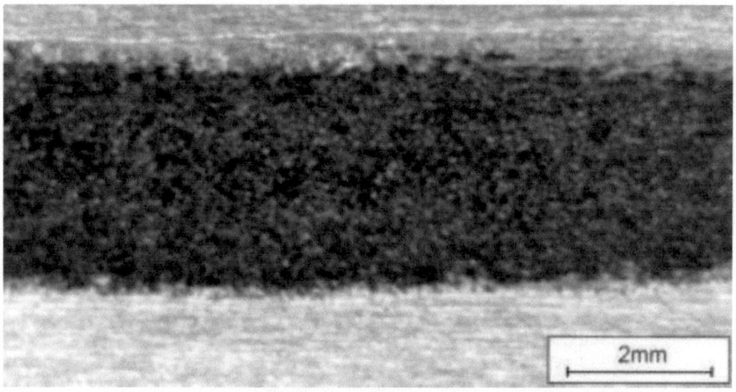

Figure 20. Surface layer of the MCMgAl3Zn1 alloy after SiC powder feeding, laser power 1.6 kW, laser scan rate 0.75 m/min

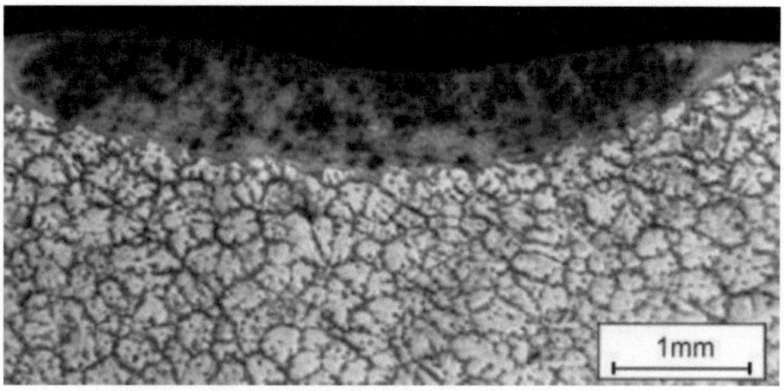

Figure 21. Surface layer of the MCMgAl3Zn1 alloy after WC powder feeding, laser power 1.6 kW, laser scan rate 0.75 m/min

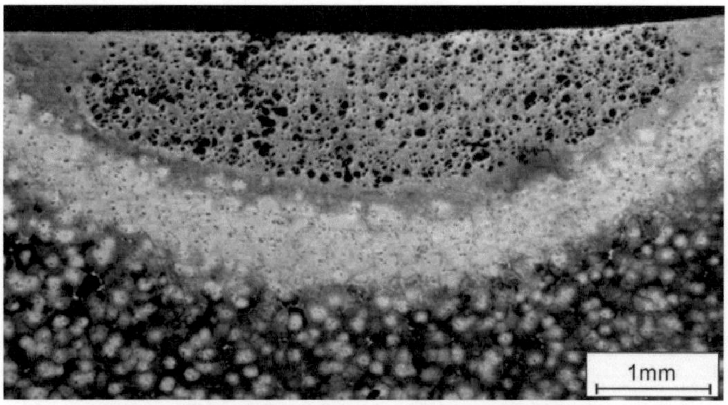

Figure 22. Surface layer of the MCMgAl9Zn1 alloy after WC powder feeding, laser power 2.0 kW, laser scan rate 0.75 m/min

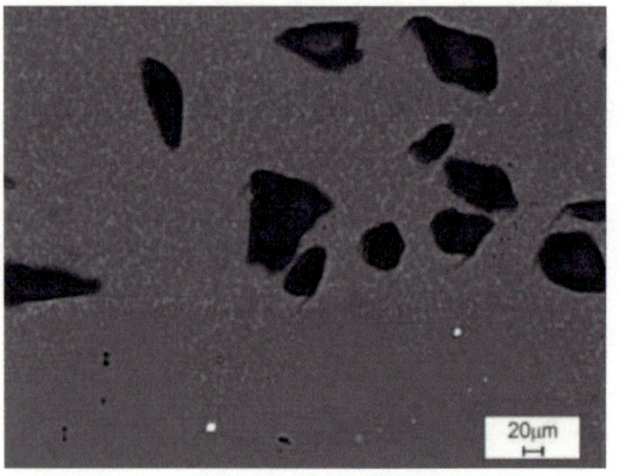

Figure 23. Central zone of the surface layer of the MCMgAl6Zn1 alloy after Al_2O_3 powder alloying, laser power 2.0 kW, laser scan rate 0.50 m/min

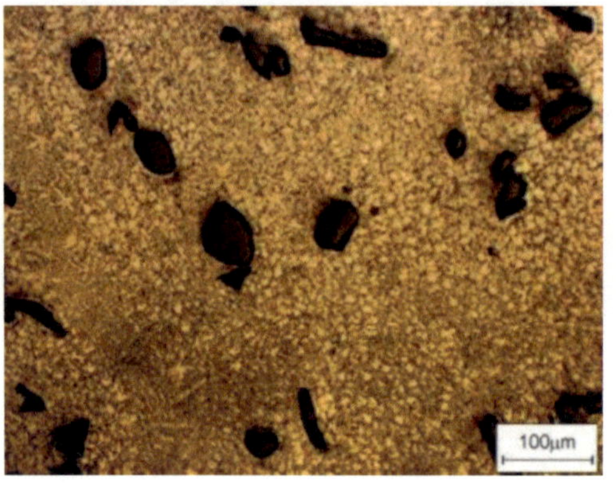

Figure 24. Central zone of the surface layer of the MCMgAl6Zn1 alloy after SiC powder alloying, laser power 2.0 kW, laser scan rate 0.50 m/min

These zones, depending on the used laser power and the applied ceramic powder are of varying thickness and shape (Fig. 25). It is possible to obtain such an effect of significant refinement of the grains only because of a fast heat transport from the remelting liquid metal lake through magnesium substrate of high thermal capacity and very good thermal conductivity, which results with the increase of grain boundaries amount representing a solid barrier for the dislocations movement and therefore reinforcement of the material.

The structure of the solidified material after laser treatment is characterised, like mentioned before, with a zone like structure with diversified morphology related to the crystallisation of magnesium alloys, containing particle of the alloying powders (Figs. 26, 27), serving also like dislocation barrier. Multiple change of crystal growth direction has been observed for these areas. In the area located on the boundary between the solid and liquid phase, minor dendrites occur with their main axes oriented along to the heat transport directions.

http://dx.doi.org/10.5772/57587

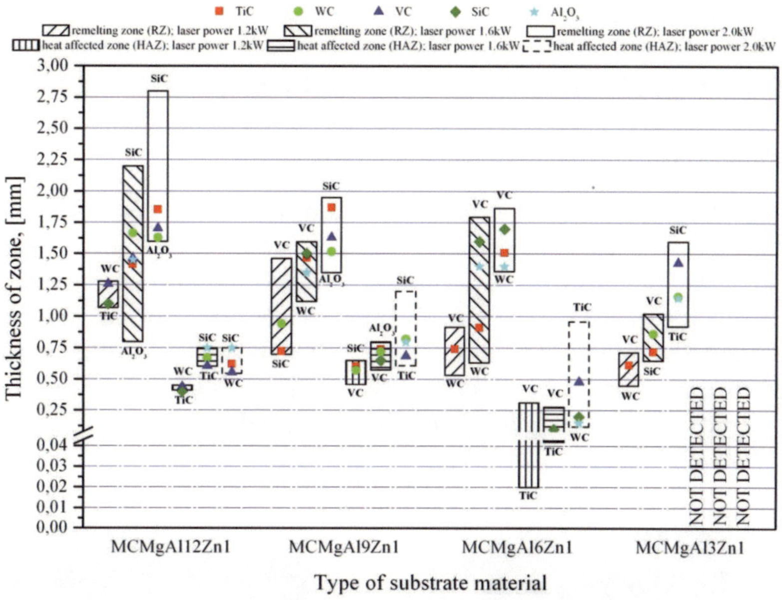

Figure 25. Influence of laser power on thickness of the remelted zone RZ, heat affected zone HAZ and the surface layer of cast magnesium alloys after laser feeding

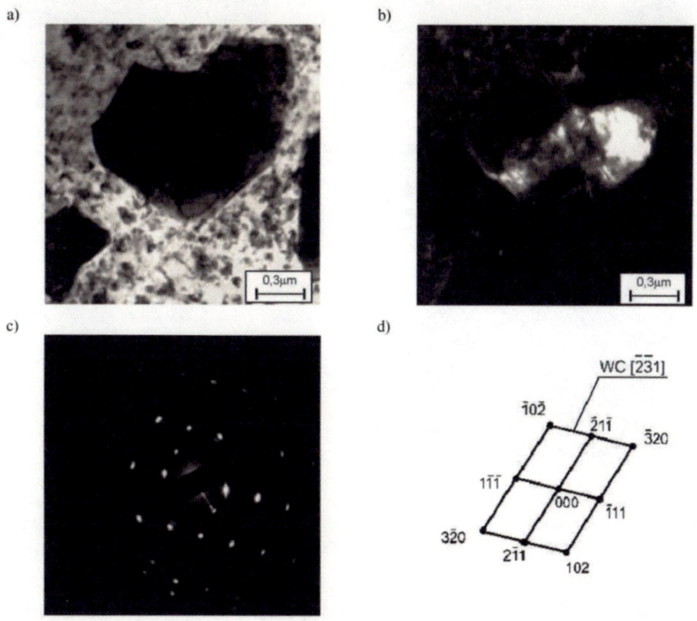

Figure 26. Structure of the MCMgAl6Zn1 magnesium cast alloy after laser feeding with WC powder, laser power 1.6 kW, a) bright field, b) dark field image from the [0-13] reflection c) diffraction pattern from the particle in Fig. a, d) solution of the diffraction pattern from Fig. c

The obtained results carried out by the team of authors concerning laser surface treatment of aluminium alloys reveal, that the highest hardness occurs in case of ZrO_2 powder alloying. In the initial planning step of investigations, there were choose specific research topics, using the innovative science-heuristic analysis for selection of the investigated material (substrate) like the Al-Si-Cu alloys based on the hardness results (Fig. 28).

Microstructure investigations of the aluminium alloys fed with WC, SiC Al_2O_3 and ZrO_2 ceramic powders (Figs. 29-38) were performed light microscopy. Like in the previous case also here the roughness of the surface increases (Figs. 29, 30). It was also found out, that in case of ZrO_2 powder alloying (Fig. 33) there was obtained a layer film, whereas in case of WC (Fig. 34) there are present particles in the surface layer; in both cases there are no pores or cracks in the produced coating or any defects and failures occurs in this layer. For the WC powder the particles are partially present on the bottom of the remelted zone.

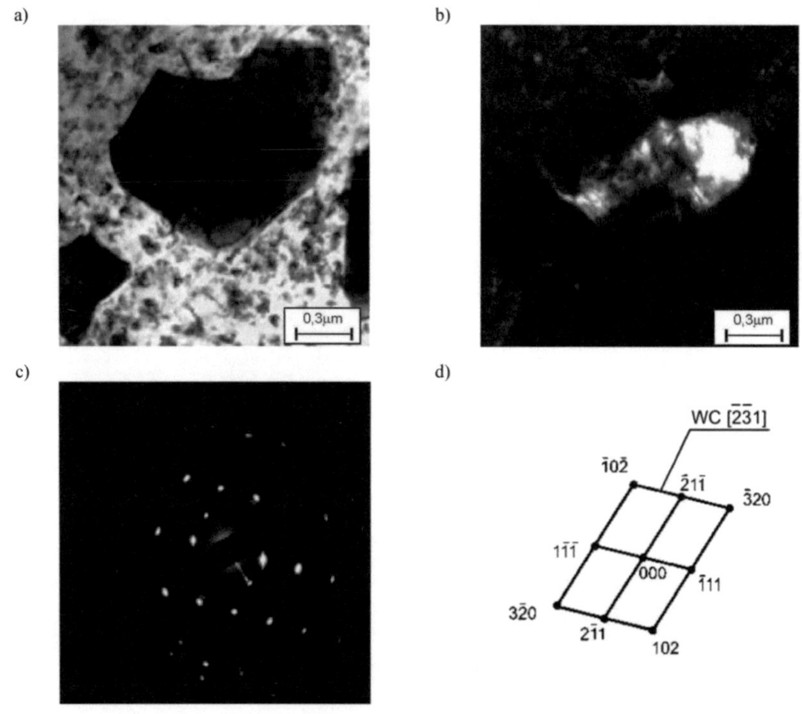

Figure 27. Structure of the MCMgAl6Zn1 magnesium cast alloy after laser feeding with WC powder, laser power 2.0 kW, a) bright field, b) diffraction pattern from the particle in Fig. a, c) solution of the diffraction pattern from Fig. b

Further investigations should revealed the exact morphology and nature of these sublayers after alloying with different ceramic powders, carried out under different process parameters. Moreover there can be recognised, that the obtained surface are characterised by a well formed structure without any breaks or defects, they are uniformly horizontally deposited in the substrate surface. Occasionally occurred discontinuities of the layer can be recognised as a

http://dx.doi.org/10.5772/57587

product of the heat transfer processes and may be neutralised by properly adjusted powder quality and powder feed rate. However in some cases the surface can reveal a very irregular shape (Fig. 35) with decomposition of the fed particles (Fig. 36). It was also found that the examined layers consists of three subzones – the remelted zone, the heat influence zone with a dendritic structure and the substrate material (Figs. 37, 38).

The thickens of the powder feed depth can be determined in the range up to 2.1 mm (Fig. 39) in case of the HAZ of WC powder fed with laser power of 1.5 kW for the AlSi9Cu4 alloy and 1.9 mm for the AlSi9Cu alloy.

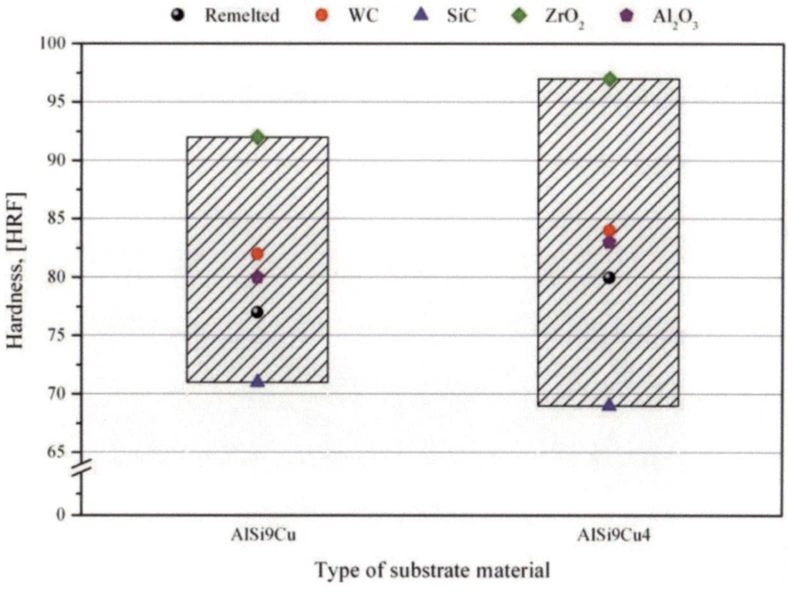

Figure 28. Hardness measurements results of cast Al-Si-Cu aluminium alloys samples, after aging and laser feeding

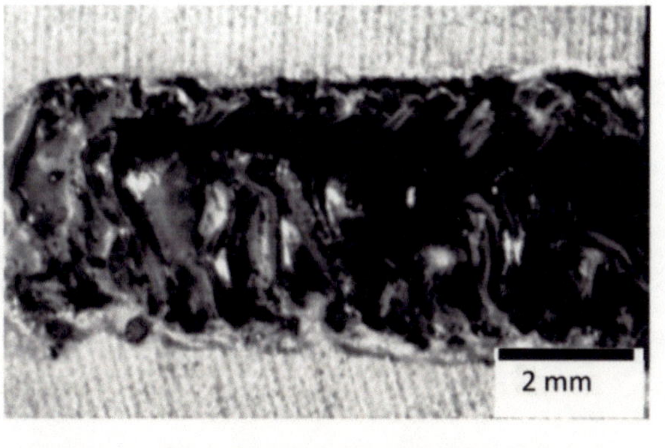

Figure 29. AlSi9Cu alloy Al_2O_3, laser power 2.0 kW, 1g/min, laser scan rate 0.5 m/min

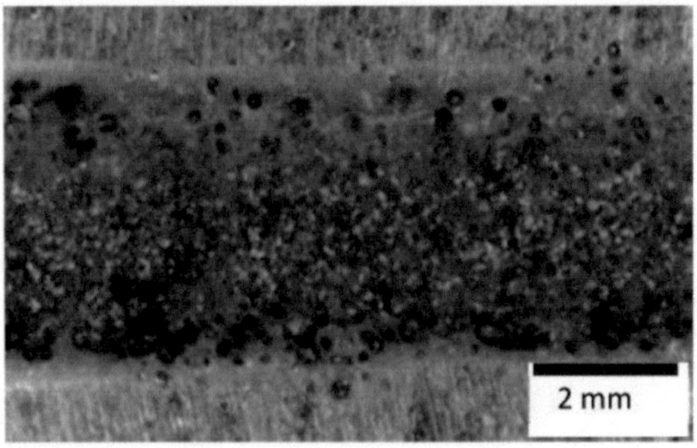

Figure 30. Surface layer of the AlSi9Cu alloyed with WC powder AlSi9Cu, laser power 1.5 kW, 1.5g/min, laser scan rate 0.25 m/min

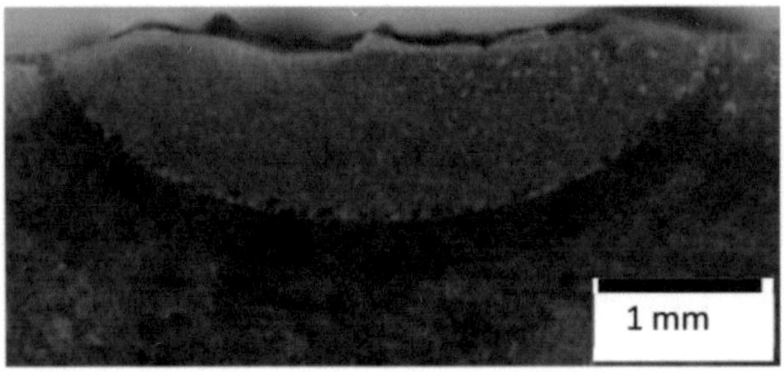

Figure 31. Cross-section of the AlSi9Cu alloy after Al_2O_3, powder feeding, laser power 2.0 kW, 1g/min, laser scan rate 0.5 m/min

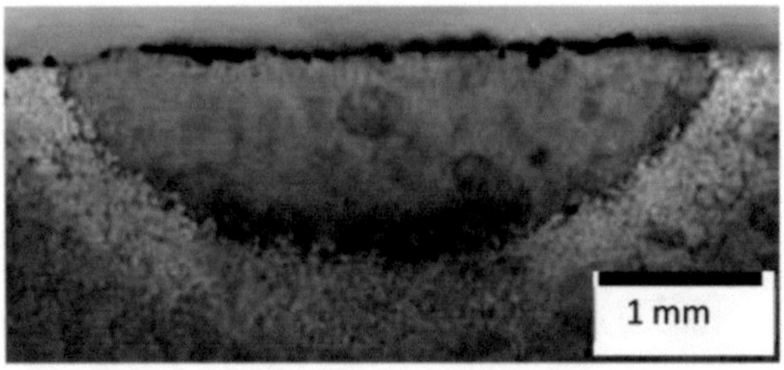

Figure 32. Cross-section of the AlSi9Cu alloy, laser power 1.5 kW, 1.5g/min, laser scan rate 0.25 m/min

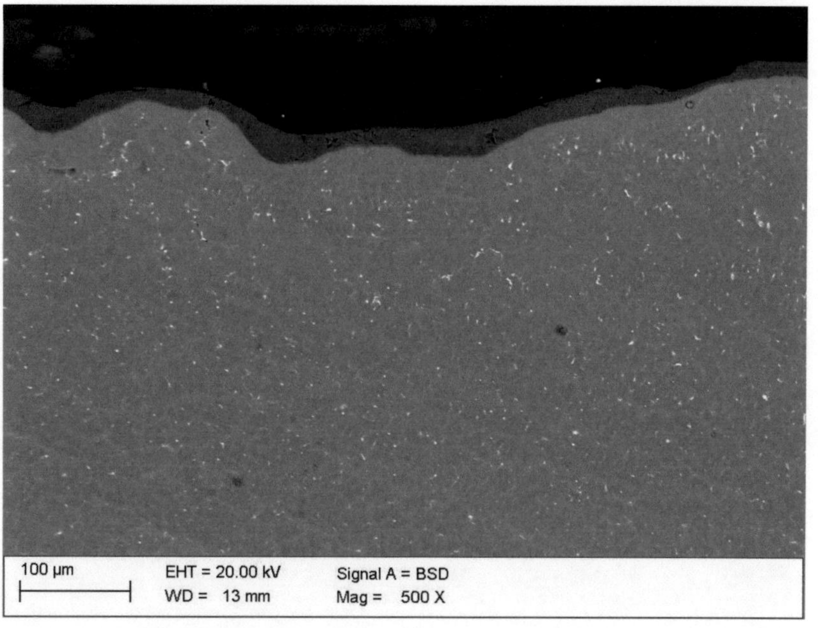

Figure 33. Surface layer of the AlSi9Cu alloy after Al_2O_3 powder feeding, laser power 1.5 kW, 1g/min, laser scan rate 0.5 m/min

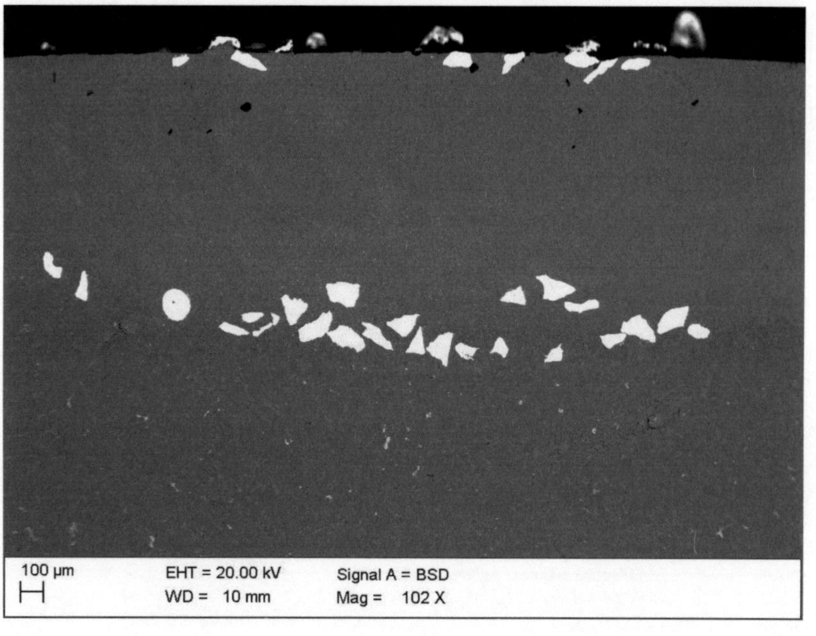

Figure 34. Surface layer of the AlSi9Cu alloy after WC powder feeding, laser power 2.0 kW, 1.5g/min, laser scan rate 0.25 m/min

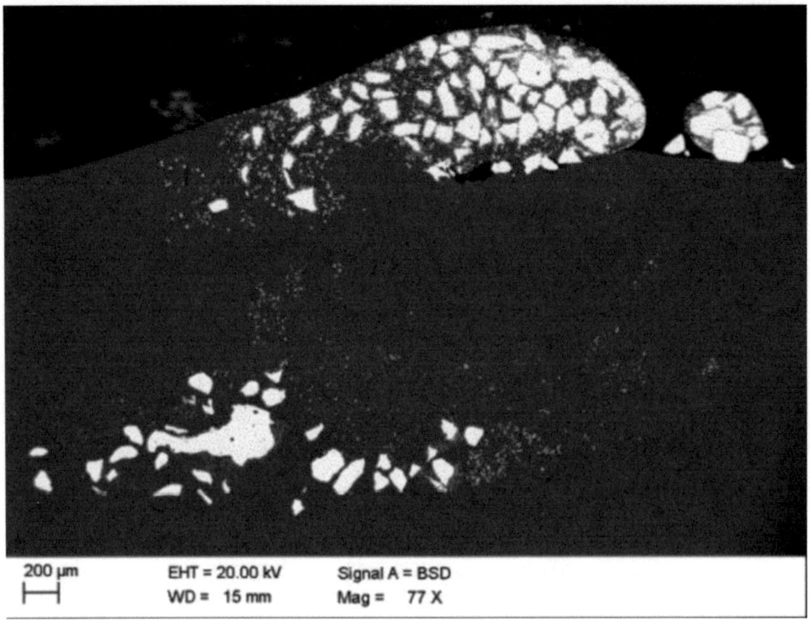

Figure 35. Surface layer of the AlSi9Cu alloy after WC powder feeding, laser power 2.0 kW, 8.0 g/min, laser scan rate 0.25 m/min

Figure 36. Surface layer of the AlSi9Cu alloy after WC powder feeding, laser power 1.5 kW, 1.5 g/min, laser scan rate 0.25 m/min

http://dx.doi.org/10.5772/57587

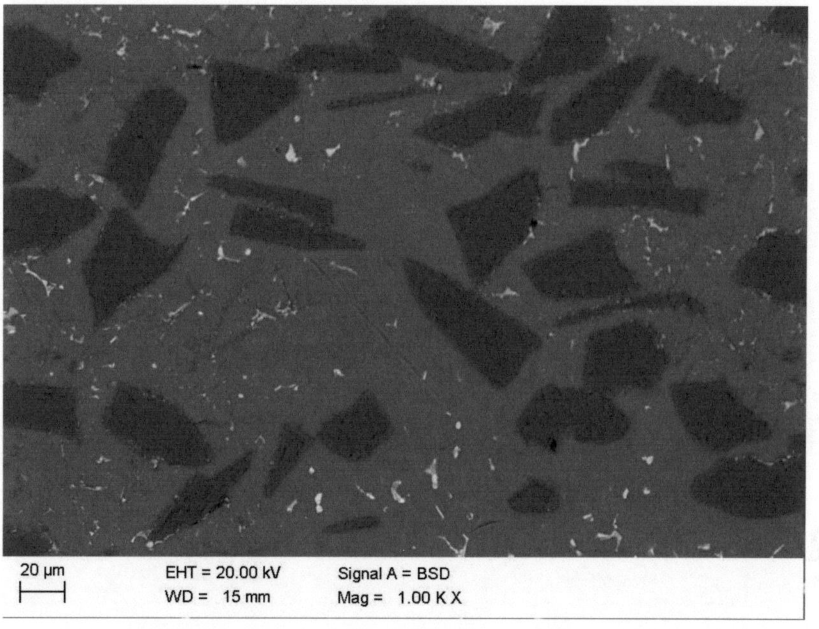

Figure 37. Structure of the surface layer of the AlSi9Cu4 alloy, fed with SiC powder, laser power 2.0 kW, 1.5 g/min, laser scan rate 0.25 m/min

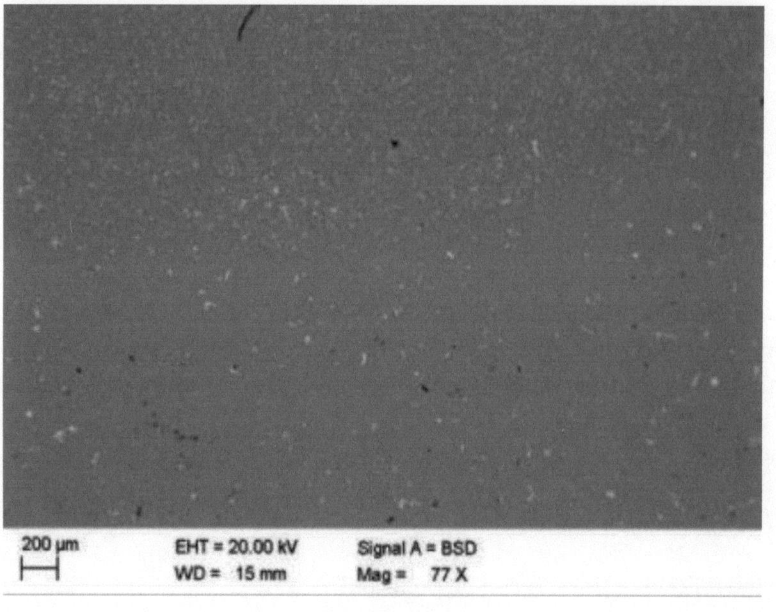

Figure 38. Transition zone between the surface layer and AlSi9Cu4 alloy substrate, fed with SiC powder, laser power 2.0 kW, 1.5 g/min, laser scan rate 0.25 m/min

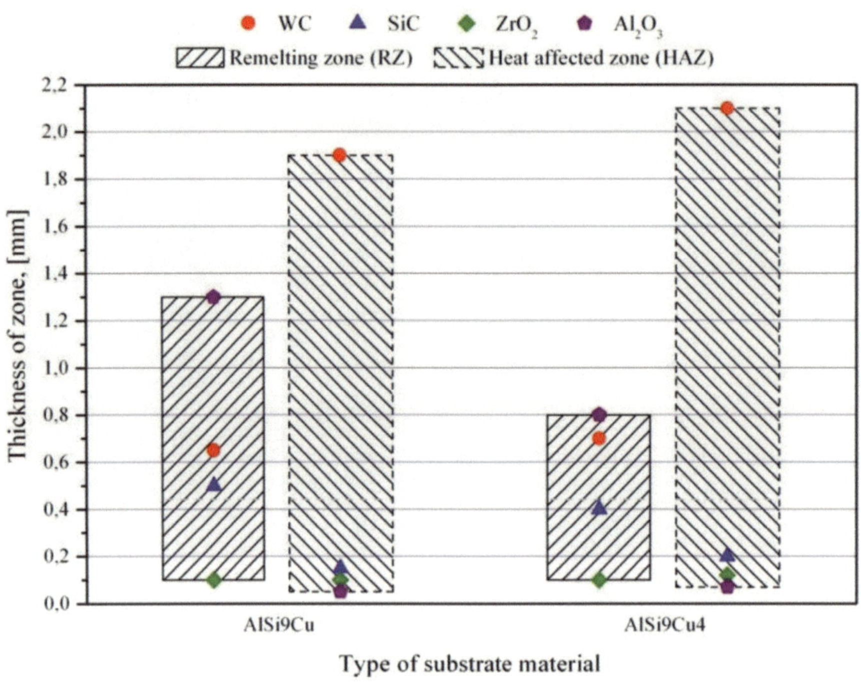

Figure 39. Influence of laser power on thickness of the remelted zone RZ, heat affected zone HAZ and the surface layer of cast aluminium alloys after laser feeding

4. Investigation results

The investigation results presented in this work reveals the positive impact of laser surface treatment on the microstructure quality and proprieties of the investigated steel as well as cast magnesium and aluminium alloys and promise further improvement in the mechanical and functional properties of the tested material, especially hardness. Laser remelting, alloying and feeding with vanadium and tungsten carbide and silicon nitride as well as aluminium oxide powders influences the refining of the structure within the entire range of laser power and the different grain size in specific zones of the investigated alloys surface layer. Considering the three groups of materials subjected to an expert evaluation using a dendrological matrix being inherent part of materials surface engineering development prediction methods, magnesium casting alloys has achieved the best position.

Concluding the following finding can be point out:

- the carried out heuristic investigations point out a very good current strategic position of the technology for the laser surface treatment of cast magnesium alloys and its possible extensive development.

The metallographic examinations carried out give grounds to state that the ceramic powder alloying or feeding process will be carried out successfully in case of the aluminium alloy substrate, the powder particles will be distributed uniformly in the investigated surface layer, and that the particular layers is without cracks and failures and tightly adhere to the cast aluminium material matrix. In this part of research following conclusion can be made:

- the laser surface treatment process gives satisfied results and allows to obtain high quality surface in most cases of the applied ceramic powders,
- the surface layer are without cracks and of a maximal thickness in the range of 2 mm,
- the surface layers consists of three zones: the remelting zone the heat influence zone and substrate material as well as sometime of an additional intermediate zone.

Steel, aluminium and magnesium are the most commonly used constructional materials after laser surface treatment, moreover there are suitable in applications as components in the automotive industry. The expected range of applications for cast magnesium alloys Mg-Al-Zn surface treatment using the high performance diode laser can be interesting especially for the automotive industry, where small product weight is required, good strength properties of components and ability to repair finished parts.

Acknowledgements

This research was financed partially within the framework of the Scientific Research Project No. 2011/01/B/ST8/06663 headed by Dr Krzysztof Labisz.

Author details

Tomasz Tański, Krzysztof Labisz and Anna Dobrzańska-Danikiewicz

Faculty of Mechanical Engineering, Silesian University of Technology, Gliwice, Poland

References

[1] T. Tański, K. Lukaszkowicz, Structure and properties of PVD coatings deposited on the aluminium alloys, Surface Engineering 28/8 (2012) 598-604.

[2] T. Tański, K. Labisz, Electron microscope investigation of PVD coated aluminium alloy surface layer, Solid State Phenomena 186 (2012) 192-197.

[3] T. Tański, L.A. Dobrzański, L. Čížek, Influence of heat treatment on structure and properties of the cast magnesium alloys, Journal of Advanced Materials Research, 15-17 (2007) 491-496

[4] L.A. Dobrzański, T. Tański, J. Trzaska, Optimization of heat treatment conditions of magnesium cast alloys, Materials Science Forum, Vols. 638-642 (2010), 1488-1493

[5] T. Tański, Characteristics of hard coatings on AZ61 magnesium alloys, Journal of Mechanical Engineering 59/3 (2013) 165-174.

[6] A.D. Dobrzańska-Danikiewicz, T. Tański, J. Domagała-Dubiel, Unique properties, development perspectives and expected applications of laser treated casting magnesium alloys, Archives of Civil and Mechanical Engineering 12 (2012) 318–326.

[7] L.A. Dobrzański, K. Labisz, E. Jonda, A. Klimpel, Comparison of the surface alloying of the 32CrMoV12-28 tool steel using TiC and WC powder, J. Mater. Process. Technol. 191\ 1/3, 321 325 (2007)

[8] L.A. Dobrzański, M. Krupiński, K. Labisz, B. Krupińska, A Grajcar, Phases and structure characteristics of the near eutectic Al-Si-Cu alloy using derivative thermo analysis, Materials Science Forum, Vols. 638-642 (2010) 475-480

[9] W. Ozgowicz, K. Labisz, Analysis of the state of the fine-dispersive precipitations in the structure of high strength steel Weldox 1300 by means of electron diffraction, Journal of Iron and Steel Research, International 18 (2011)135-142

[10] M. Piec, L.A. Dobrzański, K. Labisz, E. Jonda, A. Klimpel, Laser alloying with WC Ceramic Powder in hot work tool steel using a High Power Diode Laser (HPDL), Advanced Materials Research, vol. 15-17 (2007) 193-198

[11] M. W. Richert, B. Leszczyńska-Madej, W. Pachla, W. J. Skiba, The microstructure and properties of hydrostatically extruded polycrystalline aluminium Al99.5, Archives of Metallurgy and Materials, 57/4 (2012) 911-917.

[12] T. Tański, L.A. Dobrzański, L. Čížek, Influence of heat treatment on structure and properties of the cast magnesium alloys, Journal of Advanced Materials Research, 15-17 (2007) 491-496.

[13] L.A. Dobrzański, T. Tański, J. Trzaska, Optimization of heat treatment conditions of magnesium cast alloys, Materials Science Forum 638-642 (2010) 1488-1493.

[14] L.A. Dobrzański, R. Maniara, J. Sokolowski, W. Kasprzak, M. Krupinski, Z. Brytan, Applications of the artificial intelligence methods for modeling of the ACAlSi7Cu alloy crystallization process, Journal of Materials Processing Technology 192 (2007) pp. 582-587.

[15] L.A. Dobrzanski, W. Sitek, M. Krupinski, J. Dobrzanski, Computer aided method for evaluation of failure class of materials working in creep conditions, Journal of Materials Processing Technology 157, 2004, 102-106,

[16] 14. Dobrzański, L.A., Grajcar, A., Borek, W., Microstructure evolution of C-Mn-Si-Al-Nb high-manganese steel during the thermomechanical processing, Materials Science Forum 638 (2010) 3224-3229.

[17] L.A. Dobrzański, W. Borek, Hot-rolling of advanced high-manganese C-Mn-Si-Al steels, Materials Science Forum, Materials Science Forum 706-709 (2012) 2053-2058.

[18] T. Tański, Characteristics of hard coatings on AZ61 magnesium alloys, Journal of Mechanical Engineering 59/3 (2013) 165-174.

[19] A. Klimpel, D. Janicki, A. Lisiecki, Z. Wilk, Laser welding technologies: High power diode laser application examples, Welding International 24/9 (2010) 689-698.

[20] L.A. Dobrzański, T. Tański, J. Trzaska: Optimization of heat treatment conditions of magnesium cast alloys, Materials Science Forum, Vols. 638-642 (2010), pp 1488-1493.

[21] G. Abbas, L. Li, U. Ghazanfar, Z. Liu, Effect of high power diode laser surface melting on wear resistance of magnesium alloys, Wear 260 (2006) 175-180.

[22] D. Dube, M. Fiset, A. Couture, I. Nakatsugawa, Characterization and performance of laser melted AZ91D and AM60B, Materials Science and Engineering A 299 (2001) 38-45.

[23] Y. Jun, G.P. Sunb, S.S. Jia, Characterization and wear resistance of laser surface melting AZ91D alloy, Journal of Alloys and Compounds 455 (2008) 142-147.

[24] K.G Watkins, M.A McMahon, W.M Steen, Microstructure and corrosion properties of laser surface processed aluminium alloys: a review, Materials Science and Engineering: A 231/1–2 (1997) 55–61.

[25] K. Labisz, T. Tański, D. Janicki, HPDL energy absorption on anodised cast Al-Si-Cu alloys surfaces during remelting, Archives of Foundry Engineering 12/2 (2012) 45-48.

[26] A.D. Dobrzańska-Danikiewicz, T. Tański, J. Domagała-Dubiel, Unique properties, development perspectives and expected applications of laser treated casting magnesium alloys, Archives of Civil and Mechanical Engineering 12/3 (2012) 318-326.

Chapter 10

The Influence of Alloy Element on Magnesium for Electronic Devices Applications – A Review

Waldemar Alfredo Monteiro

Additional information is available at the end of the chapter

http://dx.doi.org/10.5772/58460

1. Introduction

This chapter presents a concise feature on past, present and the fast emerging future trends of the effective use of magnesium alloys in electronic industries. Magnesium's furthermost potential places on all types of handheld electronic devices, from mobile phones to MP3 players, cameras to smartphones, especially with the emission and interference suppression requirements of electronic devices. A lot of effort has been made to understand some properties required in electronic behavior of magnesium alloys and to have a look on their potential for further use.

To understanding some fundamental mechanism involved in any metallic material processing it is important, for example do not stop thinking about the atomic structure, chemical bonding energy and also electron distribution of the element magnesium (Table 1).

Element	Atomic number	Electron configuration
Magnesium	12	$1s^2 2s^2 2p^6 3s^2$

Table 1. Electron configurations of Magnesium

The accurate knowledge of electron configuration is particularly related with elastic properties of phases in magnesium alloy (elastic constants, the bulk moduli, shear moduli, Young's moduli and Poisson's ratio). Many of these phases are ductile. Revision involving the states density (DOS), the number of occupied orbitals (Mulliken electron occupation number) and charge density difference support the understanding of the mechanism of structural stability and mechanical properties [1].

Magnesium is the lightest of all metals used as the basis for structural alloys. The requirement to reduce the weight of automobile components as a result in part of the introduction of legislation limiting emission has triggered renewed interest in magnesium. A wider use of magnesium base alloys necessitates several parallel programs. These can be classified as alloy development, process development, improvement and design considerations [2].

Magnesium alloys are employed as a replacement for plastics when require adequate thermal conductivity properties. This makes magnesium alloys a better choice in electronic appliances to dissipate heat generated by electronic circuits. These alloys are utilized within TVs, LCDs and PC casings. There are many other better magnesium alloy properties. Magnesium alloys are less resistant to cutting, which makes them easier for rapid machining. The Mg alloys are superior at electromagnetic interference (EMI) shielding and often used in casings for mobile phones. Magnesium alloys are easily recyclable, ductile and can absorb vibration well. Magnesium alloys also have excellent resistance to corrosion. The alloys resist dents and are less likely to give dents to other metals [3-5].

Magnesium is the perfect material for applications where weight saving is a precedence, as it has the lowest density of all structural metals. Almost as light in weight as plastic, magnesium has the advantage of greater strength and rigidity, along with inherent EMI shielding and full recyclability. The magnesium benefits are: the lightest of all structural metals; excellent stiffness & strength-to-weight ratio; exceptional EMI shielding properties; high electrical and thermal conductivity; withstands high operating temperatures; cost effective versus many engineered thermoplastic materials; high dimensional accuracy and stability; exceptional thin-wall capability; good environmental corrosion resistance; good finishing characteristics; fully recyclable, all of these, ideal for electronic devices [6-8].

2. Elements in magnesium

There are several magnesium compositions which are currently being explored. Some of the alloys are at more developed research stages than others, and each alloy has been tailored for specific applications as previously mentioned. Pure magnesium has been used, as well as other elements including Zn, Mn, Al, Ca, Li, Zr, Y, and rare earth metals (RE). These within the Mg matrix create different mechanical and physical properties. If the alloying element can accomplish the metallurgical principle of developing solid solution, then solid solution strengthening can be achieved (Table 2).

Solid Solutions are generally stronger than pure metals due to the enhanced crystal structure and the introduction of crystalline defects (for example, dislocations) inside the matrix, but generally, with lower electric conductivity compared to the pure metal. This process is known as solid-solution strengthening. The formation of inter-matrix phases improves the strength of the alloy and is referred to as dispersion strengthening. This is the common practice when forming an alloy. Typically, one metal will have larger atoms relative to other constituents within the material and form a ductile phase (matrix) that consists of the major volume for the

http://dx.doi.org/10.5772/58460

Alloy description	Alloying additions	Basic Uses	Properties and applications
AZ91	9.0%Al, 0.7%Zn, 0.2%Mn	General casting alloy	Good castability, good mechanical properties at T<150 ºC
AZ31	3.0%Al, 1.0%Zn, 0.13%Mn	Wrought magnesium products	Good extrusion alloy
AM60	6.0%Al, 0.15%Mn	High pressure die-casting alloy	Better toughness and ductility than AZ91. Automotive structural applications
AM50	Mg-Al system	General casting alloy	Good strength, ductility, energy absorption properties and castability
AM20	Mg-Al system	Casting alloy	High ductility, toughness, poor die-castability
AE42	Mg-4at%Al-2at %rare earths	General casting alloy	Low castability, good creep behavior
AE44	Mg-Al-rare earth (RE) system	General casting alloy	Better creep behavior and castability than AE42
ZE41	4.2%Zn, 1.2%RE, 0.7%Zr	Casting alloy	RE addition improves creep strength at superior temperatures. Pressure tight.
AS21	Mg-Al-Si system	Casting alloy	For use at temperatures in excess of 120ºC
AS41	4.2%Al, 1.0%Si	General casting alloy	Better creep resistance than AZ91 at elevated temperatures, lower strength
AJ62	Mg-Al-Sr system	High pressure die-casting (HPDC)	Superior castability, corrosion resistance and creep behavior; good thermal and mechanical strength
MRI 153M	Mg-Al-Ca-Sr system	Casting alloy	For high temperatures applications up to 150 ºC
MRI230D	Mg-Al-Ca-Sr system	Casting alloy	For high temperatures applications up to 190 ºC

Table 2. Commercial magnesium alloys and their applications [9]

alloy. The second metal added to the alloy consists of smaller atoms that are usually both stronger and harder relative to the mechanical properties [10 – 15].

When two metals are mixed together to obtaining an alloy, the resulting material is dispersion strengthened (exist grain boundaries between two or more phases). When dispersion strengthening occurs, micro-scaled precipitates usually form within the grain boundaries, which further strengthen the material as the precipitates prevent slipping of the dislocations or other crystal defects within the grain or phase. Such action induced by heat treatment within the manufacturing process is referred to as precipitation hardening [10 – 15].

Refined magnesium almost always has trace amounts of other impurity elements. These impurities result from the natural composition of magnesium found within the earth, as well as the casting and refining processes used. The degree of impurities after the refining process is dependent upon the efficiency of the refining process itself. Elements currently found within the magnesium include copper (Cu), beryllium (Be), nickel (Ni), and iron (Fe). Standards for characteristic element inclusion in Magnesium are: 4 ppm Be, 100–300 ppm Cu, 35–50 ppm Fe, and 20–50 ppm Ni (weight percent). Other elements may be existent and are referred to as general alloying elements. Always the amount of these impurities should be controlled for any application, especially if pure magnesium is desired [13, 15 – 19].

Magnesium-zirconium-aluminum alloys are utilized in relatively low volume applications where they are processed by sand or lost-wax casting, or wrought products by extrusion or forging. The Mg-Al-Zn group of alloys is the most common alloying elements for room temperature applications. Cerium and zirconium (without aluminium) are used for elevated temperatures and form the Mg-Zn-Zr group (Table 2). Cerium is added to improve strength at the temperatures of 260°C to 370°C. Aluminium is the most effective ingredient in improving results. As little as 2% to 10% aluminium with minor additions of zinc and manganese increases strength and hardness, at the expense of less ductility, without harming weldability and making the alloy receptive to heat treatment [9].

This low density of magnesium and alloys has stimulated the use in helicopter parts, auto parts and portable electronic appliance parts. Magnesium alloys are utilized in many engineering applications where having light weight is a significant advantage. Magnesium alloys have always been attractive to designers due to their low density, only two thirds that of aluminum. This has been a major factor in the widespread use of magnesium alloy castings and wrought products [20-22].

A further requirement in recent years due to superior corrosion performance and dramatic improvements have been demonstrated for new magnesium alloys. Ultra-light magnesium alloys with excellent specific properties potentially met the present demands for lighter and reliable construction. There is a fast emerging trend towards using the lightweight materials and structures principally for electronic devices. The second principal magnesium application field are in electronic devices; see as example in Table 3 for electronic rain sensor housing.

Magnesium Alloy	AZ91D	AM60Bt
Alloy Element (%)		
Aluminum	8.3-9.7	5.5-6.5
Manganese	0.15-0.5	0.24-0.6
Zinc	0.35-1.0	0.22 (max)
Silicon (max)	0.1	0.1
Copper (max)	0.03	0.01
Nickel (max)	0.002	0.002
Iron (max)	0.005	0.005
Other Metallic (max)	0.02	0.02
Magnesium	remainder	Remainder

Table 3. Magnesium Alloys for Electronic Rain Sensor Housing

http://dx.doi.org/10.5772/58460

The AZ91D is a high-purity alloy which has excellent corrosion resistance. As mentioned before, due to low weight, good mechanical and electrical properties, magnesium is widely used for manufacturing of mobile phones, laptop computers, cameras, and other electronic components. Since its too low mechanical strength, pure magnesium must be alloyed with other elements, which confer improved properties. The tendency has been for an increase in the use of magnesium die castings and examples are computer housings and mobile telephone cases where lightness, suitability for thin wall casting and the characteristic of electromagnetic shielding are particular advantages [10, 12, 15, 16].

Magnesium alloys containing more than 1.5% Al are susceptible to stress corrosion and must be stress relieved after welding. Iron, copper and nickel are considered impurities to be limited because they degrade the corrosion resistance of magnesium alloys. Zinc combined with aluminium overcomes detrimental corrosive effects of iron and nickel impurities that may be present in magnesium alloys. The higher the Zn content (over 1%) the higher the hot shortness, causing weld cracking. Manganese improves yield strength and the saltwater resistance of magnesium alloys [10, 12, 15, 16].

Particularly for electronic devices there is alertness due to the some disadvantages of Mg alloys: high reactivity in the molten state, galvanic corrosion resistance, fire hazard, inferior fatigue and creep. The design of the Mg alloy parts is important for adequate drainage, to prevent the accumulation of corrosive substances, such as water/moisture. Iron, nickel and copper reduce the corrosion resistance of Mg alloys. Thus, during the processing of hot/molten Mg alloys, the metal must be shielded by inert gas or flux to overcome fire risk. In machining process of Mg alloys, the fire hazard/risk can be eliminated by avoiding fine cuts, dull tools, and high speeds, using proper tool design to avoid heat buildup, avoiding the accumulation of chips and dust on machines and cloths, and using coolants [10, 12, 15, 16].

Substantial research is still needed on magnesium processing, alloy development, joining, surface treatment, corrosion resistance and mechanical properties improvement to achieve future goals in all requests areas and actually, the amount of greenhouse gases [17]. Production and application technologies must be cost effective for magnesium alloys to make magnesium alloys an economically viable alternative for the automotive industry. For example the automobile contained around 20 kg of magnesium in powertrain and during its peak production consumption of magnesium reached 42,000 ton per year [6].

The global magnesium consumption is projected to rise in 2015 more than 1.3 Mt. Its single largest use, die-cast alloys for the automotive industry, is expected to show the fastest rate of about 10%p.a., underpinned largely by Chinese vehicle production. The use of magnesium cast alloys in computers, communications and consumer electronics, principally for injection molded housings, and uses for the desulphurization of steel, is also expected to grow strongly [23].

The major step for improving the corrosion resistance of magnesium alloys was the introduction of high purity alloys. Alloying can further improve the general corrosion behavior, but it does not change galvanic corrosion problems if magnesium is in contact with another metal and an electrolyte. The galvanic corrosion problem can only be solved by proper coating systems. Different coating methods are used to increase the corrosion resistance of magnesium alloys. Problems with contact corrosion can be minimized, on the one hand, by useful measures and, on the other hand, by an appropriate choice of material couple or the use of protective coatings. Aluminum coating is also used for magnesium alloys using aluminum vapor deposition and finish treatment by resin coating. The Al coating provides high corrosion resistance for magnesium alloys with various metallic sheens [7, 13, 15, 16].

3. The background for improvement — Electronic applications

Lightweight, strong and durable, magnesium alloy parts are expanding their sphere of influence into professional grade high-end appliance and electronics products. Manufacturers who are looking for ways to improve quality and reliability are fine tuning product capabilities by adding the precision and versatility of magnesium components. Magnesium is chosen for top of-the-line products that offer superior staying power, and rugged structural and mechanical properties that are unmatched by other materials. Magnesium's inherent ability to shield internal technology from electromagnetic interference and radio frequency interference (EMI/RFI) make it the ideal light metal for housing electronics.

Stand mixer has long been considered the industry standard in countertop kitchen appliances that chefs, cooks and bakers everywhere count on to mix and blend their favorite recipes. The mixer's direct drive transmission with all-metal construction is housed in magnesium. Every part of the stand mixer, from its magnesium housed transmission, to its professional-level motor, to its spiral dough hook is equipped for high performance. Product engineers,(for example, from *KitchenAid, Greenville, Ohio, USA*), selected magnesium for the stand mixer's transmission housing because the magnesium housing withstands higher temperatures and may be molded without any secondary operations, without machining for the bearing pockets and locator pins. The mixer's designers also cite magnesium as a very stable material that offers excellent part-to-part repeatability. A major benefit for the heavy-duty stand mixer is that the magnesium transmission housing may be subjected to higher loads and temperatures than the previously used thermostat plastic.

For this type of equipment, in a high-load and high-temperature application, magnesium was chosen over cast aluminum because of its ability to be molded, since cast aluminum or cast zinc would require secondary machining.

Studies have shown that one of the most important factors in determining corrosion of a magnesium implant is dependent on the purity of the material. It has been well noted that Fe, Ni, and Cu are three critical corrosion elements, which when present within magnesium significantly increase the rate of corrosion. These metals are harmful because of their low solid-solubility limits and because they serve as active cathodic sites within the material itself. Cathodic sites are regions within the material where reduction reactions occur. These regions are essential to the formation of a corrosion cell, and are generally involved with absorbing electrons produced from oxidizing materials (anodic site). At cathodic sites, nearly no material serving as the cathode is lost [13, 15, 16].

http://dx.doi.org/10.5772/58460

The solid-solubility limit of a material is the extent to which an alloying element will dissolve in base materials without forming a different phase. Generally, the higher the solid solubility limit the more likely the alloying element can homogenously disperse within a material. Whereas, low solid-solubility limits will more readily form separate phases within the material. For example, an alloy of Mg-Cu would more likely be separated into two different phases, a copper phase and a magnesium phase. This can be seen in classical cases of pitting corrosion formation.

Magnesium materials exposed to atmospheric conditions will develop a thin gray layer on its surface, which is partially protective. Ions in general are considered to be corrosive if they have the ability to breakdown the protective layer on the surface of magnesium, such as chlorides and oxidizing salts. Passivating elements like chromates, vanadates, and phosphates have been used to retard corrosion because of their ability to form a passive surface film [10-13]

Magnesium is a metal that readily corrodes in the presence of water. Atmospheric conditions contain certain levels of water content in the form of humidity, and the corrosion of magnesium alloys increases with relative humidity. The corrosion of magnesium increases significantly when exposed to pure water. This is especially true to evaluate the corrosion behavior of implantable medical devices manufactured from magnesium based materials. There are several different types of alloying elements used in magnesium based materials in an attempt to control their corrosion properties and feasibility for biomaterial implant. Elements like Mn, Cu, Al, Ca, Zr, Gd, and Zn have all been explored [11-14].

Magnesium alloy AZ91D is used, for example, by Phillips Magnesium Injection Molding, Eau Claire, Wisconsin, to form 15 different parts (power input mount housing; manifold cover; docking station battery door; RAM access, CMOS, PCMCIA, DVD, computer battery and hard drive doors; bezel for display screen frame and cover for back of display screen; the computer chassis and cover; and the docking station chassis and cover) used in the maintenance support device laptop, made by VT Miltope, Hope Hull, Alabama, has high-performance computing.

Magnesium's advantages include high strength, stiffness, durability, and superior impact resistance, making the laptop parts 20 times stronger than typical thermoplastics. The magnesium alloy housings and enclosures provide effective EMI shielding without using fillers, with an applied conversion coating. Magnesium's EMI shielding ability is critical during military field operations, since the magnesium parts protect the laptop from radiated and conducted emissions, electromagnetic pulses and radiation hazards, withstanding extreme temperatures, solar radiation, shock, transportation vibration, altitude, rain, humidity, sand, dust, and salt fog [24-26].

Some especial consumer camcorder is housed in a magnesium alloy frame, making the handheld unit much lighter and easier to hold. The magnesium frame enclosure houses a host of high-tech electronics, and includes features such as a backlit 3D button enabling easy switching from 2D to 3D mode, 5 times optical zoom, the ability to access the battery port and SD card slot while mounted on a tripod, and sports a mottled finish for an easier grip and high-end look.

Another good example is a professional video camera – magnesium and aluminum co-star as die-cast precision components for the out-of-studio camera's base, chassis, outer panels and internal sub-assemblies. The electronic engineers required lightweight, high-strength components to protect the advanced CCD image sensors, digital video processing and digital control systems inside. Magnesium industries was chosen these demands by die casting the portable video camera's housing components using magnesium AZ91D and aluminum 380 alloys for the camera's chassis, base, case panels and related parts [25-26].

Magnesium castings form the front-end chassis, frame, handle, and right and left cover panels and base, while aluminum takes the back-end role of left, right and top cover panels and back frame. An ingenious internal magnesium/front and aluminum/rear stop block system enables a miniature camera to slide into and out of the main video camera's shell, allowing versatile options for televising in-studio or remotely.

The Integrated Imaging Capsule allows the virtual plug-in mini-camera to nest inside the studio camera head or the portable camera head, delivering equally precise advanced imaging from each camera version. The die-cast rigid magnesium chassis and case panels provide the internal electronics with built-in EMI/RFI shielding. Inside, the chassis is center-gated, with walls cast to 1.016 mm. Outside panels are cast to net shape and some parts receive secondary machining, including: CNC hole drilling and tapping; minimum vibratory noise; hand cleaning; coating; final powder coat or wet paint [25-26].

Since the 90 years the electronics industries try to obtain a new type of clean magnesium alloy known as eco-magnesium alloy [27-28]. Finally in 2012, one of these real advances was developed for LG by the Korea Institute of Industrial Technology (KITECH) with funding from the Korean government, for parts being produced for all of its mobile phones.

In special LG trusts on magnesium parts for its mobile phones due to its lightness and hardness, but sought out a way to mass-produce the magnesium parts using an environmentally responsible method. The die casting process used to produce Eco-Magnesium components virtually eliminates use of damaging sulfur-hexafluoride (SF6) cover gases. As a result, LG plans a reduction in greenhouse gas emissions by a factor of approximately 24,000 during the die casting process without affecting product quality [29, 30].

Technical texts of LG indicate that is considering the use of Eco-Magnesium as soon as possible to other portable devices, such as laptops and tablets. The potential environmental benefits for expanding Eco-Magnesium use to additional electronic devices are substantial, given the company's projections for massive reductions in greenhouse gas emissions from their mobile phone production alone. Globally, manufacturers in key industries are seeking environmentally sound ways to produce and form magnesium alloys. Magnesium's spectrum of sustainable product possibilities is expected to broaden and grow, as new forming and processing technologies enable the use of magnesium parts in a host of new high-performance products [29, 30].

Within the last fifteen years several heavy magnesium parts have been assembled in passenger cars, such as gearbox housings and crankcases. Reasons are the new heat resistant alloys and the growing duty of automotive constructors to reduce the vehicles weight and in that way CO_2 emission (regulation). The rising quantity of magnesium in automobiles will result in an increasing amount of post-consumer scrap. So far magnesium recycling for magnesium alloy production is done only for clean scrap. Clean in this regard means "free from impurities", contaminations and sorted according to a chemical composition or a single phase alloy [25-28].

For automotive post-consumer scrap the materials quality strongly depends on the way end-of-life vehicles are treated. Up to date automotive manufacturers have no uniform recycling system for end-of-life vehicles. By proper dismantling of old cars, the bigger magnesium components could be collected and possibly treated similar to class 1 scrap.

So far the biggest part of old cars has been shredded and it is probable that this practice will be continued for reasons of economy. The magnesium fraction can be segregated from the non-magnetic shredder fraction via sink float separation and eddy current separation. A further separation according to chemical composition is claimed to be feasible. Still another problem is unsolved. Due to coatings on magnesium components or aluminium alloy contaminations it is expected that magnesium post-consumer scrap will be enriched with Cu, Fe and Ni during remelting [30].

Eco-conscious design innovations require materials that support and facilitate optimal energy efficiency and longer product life cycles. Magnesium alloys achieve this while being fully reusable and recyclable. From Light Emitting Diode (LED) lighting to lighter vehicles that reduce an engine's load, to housing longer-life batteries in computers and electric personal transport, magnesium is turning into the designer's most versatile material choice to improve product functionality and develop user mobility with greater energy efficiency and environmental benefits.

A light-emitting diode (LED) is a two-lead semiconductor light source that resembles a basic pn-junction diode, except that an LED also emits light. When a LED's anode lead has a voltage that is more positive than its cathode lead by at least the LED's forward voltage drop, current flows and electrons are able to recombine with holes within the device, releasing energy in the form of photons. This effect is called electroluminescence, and the color of the light (corresponding to the energy of the photon) is determined by the energy band gap of the semiconductor [31, 32].

The light's entire housing is constructed of magnesium alloy AZ91D by AltusLumen, Hong Kong, China [31, 32]. The portable LED is a convenient rechargeable lithium-ion battery light with intelligent power management, featuring a three-dimensional hinge that adjusts to any lighting angle. The ultra-bright and efficient runs for several hours on a single charge, and may be charged by solar panel with an optional solar charger because magnesium conducts heat generated by the LED light away from the LED bulb, extending its usable life.

In addition, only magnesium provides the strength required to produce the unique feature of a three-dimensional swivel hinge. The LED light component is screwed into the magnesium housing, with upper and lower housings joined by a metal pin, force fitting upper and lower housings together. The magnesium housing is cast via injection. After parts are injected, CNC machining removes extra material. Holes are then drilled into the housing for mounting and the magnesium is powder coat painted. The pocket-size portable light follows the company's guiding sustainability principles that include energy efficiency, using renewable energy, and choosing materials that are recycled and recyclable (AZ91D).

Magnesium is one of the few materials that can be continuously recycled without degradation. Magnesium is easier to recycle and its value makes it an attractive material to recycle. Another company also chooses magnesium alloy housing for its energy efficient LED warm-white and pure-white frosted soft flood lights (LEDtronics, Inc., Torrance, California) and maintains high shock and vibration resistance.

Thermal management performed using the magnesium housing offers the best heat sinks, even better than copper. The long-life LED flood light is used for signage, architectural and landscape lighting, security, aviation, industrial equipment, medical, and theatrical lighting. The magnesium swivel hinge and foldable tripod stand make the TRI-L LED portable light a hands free, go anywhere device. The Parabolic Aluminized Reflector (PAR) 38 Series LED flood light features a magnesium alloy body in a solid-state design that facilitates energy savings of up to 85 percent compared to incandescent flood lights [33].

In the modern market, small and portable electronics are trending increasing. The demand for compact devices that can be easily transported is successful and magnesium is often a key component in meeting this demand. Many magnesium alloys being used to replace plastics are just as light, but they are much stronger and more durable.

Magnesium is also better in regards to heat transfer and dissipation as well as its ability to shield electromagnetic and radio frequency interference. Many electronics require parts or casings with multifarious shapes which are possible with magnesium. Camera, cell phone, laptop and portable media device lodgings are all common applications in addition to arms of hard drives [33, 34].

Electroplated plastics might look like metals but, because the underlying plastic substrate is thermally non-conductive, they do not feel like metals.

This simple matter of aesthetics often disappoints consumers when it is realized, when touched, that an outwardly metallic looking item clearly isn't what it was thought to be. Because it is a conductive metal, AM-lite® does have a "metallic feel" and so is more attractive to customers. AM-lite® has a much higher stiffness, yield strength and creep strength than unreinforced plastics thus allowing improved thin section light weight designs. The alloy's electrical and thermal conductivity, and EMS shielding, is also important for the design of consumer electronics such as mobile phones and laptop computers [34].

Author details

Waldemar Alfredo Monteiro[1,2]

1 Materials Science and Technology Center –Nuclear and EnergyResearch Institute, São Paulo, SP,, Brazil

2 School of Engineering – Presbyterian Mackenzie University, São Paulo, Brazil

References

[1] Pingli M, Bo Y, Zheng L, Feng W, Yang J, First-principles calculations of structural, elastic and electronic properties of AB2 type intermetallics in Mg – Zn – Ca-Cu alloy", Journal of Magnesium and Alloys 1 (2013) 256-262

[2] Mordike BL & Ebert T, Materials Science and Engineering A302 (2001) 37–45.

[3] Kojima Yo, Platform Science and Technology for Advanced Magnesium Alloys, Magnesium Alloys 2000, Materials Science Forum, Volumes 350-351), page 3 – 19, doi: 10.4028/www.scientific.net/MSF.350-351

[4] Duffy L, Materials World, Vol 4. No 3, pp. 127-30, 1996.

[5] Dunlop G.; Abbott T.; Murray M.; Bettles C.; Gibson M.; in Proceedings International Magnesium Association World Conference, Beijing, IMA 2006, p. 24 – 33

[6] http://www.magnesium-elektron.com/markets-applications.asp?ID=7, 23/02/2014.

[7] Kainer K. U., Dieringa H., Dietzel W., Hort N. and Blawert C., The Use of Magnesium Alloys; Past, Present and Future, Pekguleryuz M.O. and Mackenzie L.W.F. eds. (2006) Magnesium Technology in the Global Age 2006, Metallurgical Society of CIM, Montreal Canada. 2006 Magnesium Technology in the Global Age

[8] Gibson M. A., Bettles C. J., Murray M.T., Dunlop G.L., Cashion S.P., "AM-HP2: A New Magnesium Alloy with Improved Diecastability and Creep Strength for Powertrain Applications", 13th Magnesium Automotive and End User Seminar, Aalen, 22nd-23rd September 2005.

[9] Kulekci M. K., Magnesium and its alloys applications in automotive industry, Int. J. Adv. Manuf. Technol. (2008) 39: 851-865

[10] Eliezer D., Aghion E., Froes F.H., Magnesium Science, Technology and Applications, Advanced Performance Materials 5, 201-212 (1998) Kluwer Academic Publishers, The Netherlands.

[11] Polmear, I.J., Light alloys: metallurgy of the light metals, Third edition, 1995, Arnold, London, ISBN 0-340-63207.

[12] Smith, W.F., Structure and properties of engineering alloys, 2nd edition, 1993, McGraw-Hill, ISBN 0-70-112829-8.

[13] Kainer, K.U. (Ed), Magnesium Alloys and Technology, DMG, 2003, Wiley-VCH, ISBN 3-527-30256-5.

[14] Siobhan F, An Overview of Magnesium based Alloys for Aerospace and Automotive Applications, MSc, Rensselaer Polytechnic Institute, Hartford, CT, August, 2012.

[15] Avedesian, M.M., Baker, H., ASM SPECIALTY HANDBOOK, Magnesium and magnesium alloys, 1999, ISBN 0-87170-657-1.

[16] E. Aghion, B. Bronfin, Magnesium Alloys Development towards the 21st Century, Magnesium Alloys 2000, Materials Science Forum, Volumes 350-351), page 19 – 31, doi:10.4028/www.scientific.net/MSF.350-351

[17] Ehrenberger S., Schmid S., Friedrich H., Magnesium production and automotive applications: Life-cycle analysis focusing on greenhouse gases, 16. Magnesium Abnehmer-und Automotive Seminar 2008 in Aalen (http://www.efm-aalen.de/PDF/Ehrenberger_magnesiumx.pdf)

[18] Kubota K., Mabuchi M., Higashi K., Review Processing and mechanical properties of fine-grained magnesium alloys, Journal of Materials Science, Volume 34, Issue 10, pp 2255-2262.

[19] Labelle P., "Compilation of AJ62A Typical Alloy Properties", Noranda Inc., December 2004.

[20] Dieringa H., Bohlen J., Hort N., Letzig D., Kainer K. U., Advances in Manufacturing Processes for Magnesium Alloys, Beals, R. S., Luo, AA, Neelameggham, NR, Pekguleryuz, MO, eds (2007) Magnesium Technology 2007, Orlando, FL 2007 Magnesium Technology

[21] Avraham S., Maoz Y., Bamberger M., Structural Evolution in Mg-Al Alloys, Beals, RS, Luo, AA, Neelameggham, NR, Pekguleryuz, MO, eds (2007) Magnesium Technology 2007, Orlando, FL, 2007 Magnesium Technology

[22] Longanbach S. C., Knittel K., Lee A., Boehlert C. J, Processing-Microstructure-Property Relationships of Magnesium Alloys Containing Zr and/or B, Beals, RS, Luo, AA, Neelameggham, NR, Pekguleryuz, MO, eds (2007) Magnesium Technology 2007, Orlando, FL 2007 Magnesium Technology

[23] http://www.chinamagnesiumcorporation.com/our-business/magnesium-overview.

[24] Barker E., Li D., Sun X., and Khaleel M., Microstructure Modeling of Magnesium Alloys for Engineering Property Prediction, Mathaudhu, SN, Sillekens, WH, Neelameggham N.R., Hort N, eds (2012) TMS Magnesium Technology 2012, Orlando, FL, USA, 2012 Magnesium Technology

[25] Ulrike G. K. Wegst, Michael F. Ashby, The development and use of a methodology for the Environmentally-Conscious selection of Materials, Third Biennial World Conference on Integrated Design and Process Technology (IDPT), 6-9 July 1998, Berlin, Germany (Proceedings Volume 5, pp. 88-93)

[26] Aghion E., Bronfin B., Friedrich H. and Rubinovich Z., "The Environmental Impact of New Magnesium Alloys on the Transportation Industry", Magnesium Technology 2004, ed. A.A. Luo (Warrendale, PA: The Minerals, Metals & Materials Society, 2004), 167-172.

[27] Ninomiya R., Kamado S., Kojima Y, Morinago M., An Electronic Approach to the Prediction of the Mechanical Properties and the Design of Magnesium Alloys, Mor-

dike, BL, Kainer, KU eds (1998) Magnesium Alloys and their Applications, Hamburg, Germany, 1998 Magnesium Alloys and their Applications

[28] Su X., Li D.J., Zeng X.Q., Wang Y.X., Cheng L.F., Ding W.J., Elastic Properties and Electronic Structures of Mg-Re (Re=Sm, Y) Alloys Studied by First Principle Calculation Principle Calculation, Poole, W. J. and Kainer, K. U., eds (2012), Proceedings of the 9th International Conference on Magnesium Alloy and Their Applications, July 8-12, 2012, Vancouver, BC, Canada, 9th International Conference on Magnesium Alloys and Their Applications.

[29] http://news.cnet.com/8301-17938_105-20024756-1.html, 23/11/ 2012.

[30] http://www.lg.com/global/press-release/article/lg-to-utilize-only-cleaner-eco-magnesium-in-all-mobile-devices-by-2012.jsp, 18/06/2013.

[31] Fechner D. et al. Sci. China Ser. E-Tech Sci. Jan. 2009, vol. 52 no. 1 148-154

[32] http://www.candlepowerforums.com/vb/showthread.php?226594-AltusLumen-PAD-L-Review-Portable-Flood-Light, 23/01/2014.

[33] http://www.intlmag.org, 18/01/2014.

[34] http://www.amlite-lighting.com, 17/10/2013.